CARBONshift

CARBON

How the Twin Crises of Oil Depletion and
Climate Change Will Define the Future

edited by Thomas Homer-Dixon
with Nick Garrison

RANDOM HOUSE OF CANADA

LIBRARY AND ARCHIVES CANADA CATALOGUING IN PUBLICATION

Carbon shift : how the twin crises of oil depletion and climate change will define the
future / Thomas Homer-Dixon, ed.
Includes index.

ISBN 978-0-307-35718-2

1. Petroleum reserves—Forecasting. 2. Petroleum industry and
trade—Forecasting. 3. Global warming. 4. Climatic changes. 5. Energy
policy. 6. Carbon dioxide mitigation. I. Homer-Dixon, Thomas, 1956–

QC981.8.G56C372 2009   333.8'23211   C2008-906526-3

Design by Andrew Roberts

Printed in the United States of America

10 9 8 7 6 5 4 3 2 1

Contents

Ronald Wright
FOREWORD

Ronald Wright is an award-winning novelist, historian and essayist. He is the author of ten books, including the #1 bestsellers *Time Among the Maya* and *Stolen Continents*. His first novel, *A Scientific Romance*, won Britain's David Higham Prize for Fiction and was chosen a Book of the Year by the *Globe and Mail*, the *Sunday Times*, and the *New York Times*. His book *A Short History of Progress* (the 2004 CBC Massey Lectures) won the CBA Libris Award for Non-Fiction Book of the Year and has been published worldwide in more than a dozen languages. His most recent book, *What Is America? A Short History of the New World Order*, was a #1 national bestseller.

Civilizations are built on knowledge, population—and energy. They thrive only when a good balance is struck between these three, a balance dependent (like that of a bicycle) on motion, which is to say on growth. Human successes are always taken from the past or borrowed from the future: sooner or later the bike runs out of road. The first humans evolved by devouring the great wild beasts that once roamed all parts of the Earth. When they exhausted this primordial energy hoard at the end of the last ice age, they starved; and the humble survivors—our ancestors—became more and more dependent on plants.

Over time, early civilizations arose with the development of systematic agriculture. Through crop breeding, animal husbandry, deforestation and irrigation, they concentrated the energy of soil and seeds into the muscle power of domesticated animals and equally domesticated human beings. Towns, cities, governments and priesthoods rose like pyramids on a broadening agrarian base. Despite booms and busts along the way, humanity grew at an ever-increasing rate, especially after the crops of the Americas (such as maize and potatoes) spread around the world. By some two hundred years ago, human beings had reached the maximum number who could feed themselves by muscle power and pre-industrial machinery. That number was about one billion.

What has allowed us to soar nearly sevenfold since then was not any breakthrough in new food: all our crops are ancient; we have raised yields by tinkering, but we have developed no new staples from scratch since prehistoric times. The breakthrough was in energy—in finding new ways to use the vast stocks of fossil carbon that Nature had buried under the planet's skin long before the first mammal crawled upon it.

We tend to think of the looming energy crisis in terms of cars, factories, heating and air conditioning, but the first thing to keep in mind is that fossil fuels are feeding us. We all know that coal and oil drive the tractors, trains, trucks, ships and freezers that grow, store and move food from farm to city, nation to nation. But how many are aware that we have literally been eating oil and gas for more than a hundred years? Fossil carbon is a prime ingredient of the artificial fertilizers that have sidestepped the decline of natural fertility each time a crop is taken off a field. A two-century carbon binge has allowed mankind to fill its planet way beyond the natural carrying capacity for feckless, reckless, self-indulgent apes. If we run out of carbon or fail to find good substitutes, we are back to dung and muscle power. Billions will die.

An absolute shortage of fossil energy is still a long way off. But the amount that can be easily, cheaply and above all safely exploited is indeed running low. Because of carbon dioxide's effect on climate, an abundance of carbon fuel—

especially in its dirtier forms such as coal and tar sand—is far more dangerous than a dearth. Long before fossil fuel gets truly scarce, its consumption will overthrow the predictable weather patterns on which all farming has relied for the past ten thousand years. In short, the industrial carbon economy has turned out to be what I call a "progress trap"—a seductive and seemingly benign development which, upon reaching a certain scale, becomes a dead end.

Even if abundant sources of clean energy were to come on stream tomorrow, we would still face problems of overpopulation, overconsumption, soil erosion and the most unequal distribution of wealth and health in history. But, as the essays in this important book explore and document in different ways, a "carbon shift"—a swift transition to much cleaner energy—is our only hope of escaping the dire consequences of our runaway success.

Thomas Homer-Dixon and Nick Garrison
INTRODUCTION

Thomas Homer-Dixon was born in Victoria, B.C., and holds a
Ph.D. in political science from MIT. He is currently the CIGI
Chair of Global Systems at the Balsillie School of International
Affairs at the University of Waterloo. His #1 bestselling first
book, *The Ingenuity Gap*, won the 2001 Governor General's
Literary Award for Non-Fiction, and his most recent book,
*The Upside of Down: Catastrophe, Creativity and the Renewal of
Civilization*, was a #1 national bestseller and won the 2006
National Business Book Award.

Nick Garrison is a writer and editor, and former communications
director of a prominent environmental NGO. He lives in Toronto.

People just want to go on doing what they're doing. They want business as usual. They say, 'Oh yes, there's going to be a problem up ahead,' but they don't want to change anything.

—James Lovelock

Whenever people are given a chance to make a wish, they almost always wish for too much. The consequences are invariably unintended. Sometimes they're catastrophic.

In the days before cars and trucks and coal-fired power plants, Silenus, satyr and tutor to the Greek god Dionysus, drank immoderately and wandered into the rose garden of the king of Phrygia, where he promptly fell asleep. Recognizing his uninvited visitor, the king brought the sleeping tutor inside and entertained him for ten days of feasting and storytelling before taking him back home to Dionysus.

Dionysus was pleased to see his old tutor, and asked the king to choose his own reward for taking such good care of him. The king, whose name was Midas, asked that whatever he touched turn to gold. The wish, of course, was granted, and it was not long before the king realized it was a curse. His food turned to gold as he raised it to his lips. His daughter was transformed into cold metal as he reached out to caress her. The Midas touch deprived him of the very best things in life, and he soon begged Dionysus to take the gift back.

Taking pity on the foolish king, the god directed Midas to wash his hands in the river Pactolus in what is now Turkey.

When he did, the power to transform passed from the king to the water itself, and the Pactolus ran with magical water that turned whatever it touched to gold.

Here myth merges with history: the sands of the Pactolus were in fact laced with powdered gold, a source of great wealth to the Iron Age kingdom of Lydia. According to the ancient Greek historian Herodotus, the Lydians were the first to use metal coins as currency and the first to set up retail shops. Gold and electrum from the Pactolus gave the Lydians the power to bribe officials and influence the affairs of states as far away as mainland Greece and Egypt. And Croesus, the last king of the Lydians, is still remembered as a metaphor for wealth. His Temple of Artemis at Ephesus was one of the seven wonders of the ancient world. What had been a curse to Midas turned out to be something the Lydians could not live without.

But for us it hasn't been gold but rather carbon that has granted our wishes for wealth and power and that now threatens to transform our world into something profoundly inhospitable to human civilization.

By nearly any measure we've got what we wanted. Perhaps we've got more than we even knew we wanted. Even just a few decades ago, things that many people today take for granted were reserved only for the very privileged: frequent air travel, second homes, regular meals in restaurants and sleek imported cars. We wished for speed, mobility, headlong economic growth and an array of dazzling consumer goods. And we got it all—at least those of us in the world's rich countries got it all. Now the rest of the world wants it all too.

Within the lifetimes of people still alive today, both our economies and our day-to-day lives were marked by the constraints of the natural world. We grew food, harvested trees and minerals, and worked with our hands in fields and factories. But within that single lifetime, new technologies coupled with the astonishing creative destruction of global capitalism have changed everything. Manufacturing, shipping, retailing, marketing, financing and consuming in a global just-in-time market have transformed the world. Businesses with tens of billions of dollars in revenues—medical research and services, entertainment, sports, education, fashion and the like—span and homogenize the planet's economies and cultures. All this has brought a substantial fraction of humanity higher incomes, longer lives and a standard of living our great-grandparents would scarcely recognize.

Environmentalist James Lovelock's "business as usual" doesn't sound so bad. And by some accounts things will only get better. According to a report by Goldman Sachs, within fifty years, if the global economy continues on its current trajectory, the average income in China will be about what it is in the United States today.[1] But then, too, per capita GDP (gross domestic product) in the U.S. will be more than double what it is today. Turkey, Nigeria and Iran will have economies on par with today's G7 nations. The so-called BRIC nations (Brazil, Russia, India and China) will either challenge those in the top spots or occupy the spots themselves. If business continues as usual, according to the standard storyline, everyone should get a lot richer.

But this outcome depends on something that, with every passing month, seems less and less certain: a non-stop rise in our already staggering energy consumption. The capitalist economic machine that has been producing prosperity, innovation and (some would say) political stability since the Industrial Revolution now requires a constant flow of almost unimaginably large quantities of energy—especially energy supplied by fossil fuels. This year we'll burn eight billion metric tonnes of oil, natural gas and coal, and every year we burn more.[2]

Like the fabled sands of the Pactolus for Lydian civilization, our fossil fuels have become essential for the existence of humankind's industrial civilization. The problem that fossil fuel depletion, especially conventional oil depletion, poses for our civilization is not as simple as having the gas gauge of our car fall towards "empty" but, if we're not smart soon, the end result could be largely the same. If we're unprepared as we approach empty, the machine won't work anymore, and we'll be left stranded at the side of the road.

Without fossil fuels, and without aggressive efforts to find alternatives, much of what we take for granted will be impossible. Take the book in your hand. It's an artifact of an economy made possible by fossil energy—the bulldozers, chainsaws and logging trucks required to harvest trees and ship them to pulp mills; the mills themselves that turned trees into paper; the presses that printed this book; the trucks that shipped it; the forklifts in the warehouses that stocked it; and the lights and heat in the store that sold it.

And that's something as old-fashioned as a book. The energy costs of running the banks of Google or Facebook servers, or even of running our computers at home, are far larger than most people realize. Like printing on paper, running modern information technologies has very real environmental implications. The average computer data centre

consumes as much energy as 25,000 North American homes. While the electricity that runs our printing presses, lights our buildings and keeps our computers operating *may* come from nuclear power plants or hydroelectric dams (depending on where we live), it's more likely to come from fossil fuels like natural gas or coal. Here in North America, about 33 percent of our electricity comes from hydroelectric and nuclear facilities—the rest from fossil fuel.

Our challenge is particularly acute in the case of conventional oil—that is, oil found in liquid form in reasonably accessible locations on land or near offshore. Conventional oil is fantastically useful. It's easy to ship, easy to trade and packed with energy. Our highly mobile and highly productive global economy literally runs on the stuff. Oil and natural gas have propelled the quadrupling of the world's human population over the past century, largely by powering the tractors and irrigation pumps and by providing the chemical fertilizer that have quadrupled agricultural yields. (In the last century or so, average energy inputs per hectare of Earth's agricultural land have soared eightyfold.)[3] We have converted, literally, petroleum into food and food into billions of people. And we have been able to do this because oil and gas have been ridiculously cheap.

Even when the price of a barrel of oil soared to record levels in mid-2008, oil-based energy was still an extraordinary bargain: three spoonfuls of crude oil contain as much energy as eight hours of human labour yet cost a tiny fraction of this labour at market prices. The reason is that not only is energy, especially oil, undervalued—even when oil was $150 a barrel—its economic role is only beginning to be understood.

A considerable portion of the twentieth century's economic thought was devoted to the often-ideological debate between those who thought capital the most important input to a thriving economy and those who made the case for labour. Often overlooked was the role of energy in economic growth. In 1956, economist (and later Nobel Prize winner) Robert Solow argued that, because about 70 percent of the cost of producing something went to pay for labour and about 30 percent went to pay for capital, a 1 percent rise in the contribution of labour (say, longer hours) would result in 0.7 percent extra economic growth, while a 1 percent rise in the contribution of capital (say, investment in new factories) would produce 0.3 percent growth.[4]

But Solow recognized that something was clearly amiss with this calculation, since his model did not predict the real GDP growth in the United States and other industrialized countries. Growth was invariably much higher than he predicted, and the difference—that proportion of output growth that could not be explained by the measured inputs—came to be called the "Solow residual." But as economist Reiner Kümmel later pointed out, this residual often explains more than Solow's theory itself.[5]

After Solow's work, it had been widely assumed that the "something else" that combines with labour and capital to produce economic growth must be technology. But Kümmel showed it must be energy. He realized that the Solow model accounted for energy only on the basis of what it costs, rather than what it produces. Because energy costs about 5 percent of GDP, increasing energy inputs 1 percent, according to Solow's reasoning, would expand GDP by only 0.05 percent.

But by counting energy inputs in terms of joules rather than dollars, Kümmel not only produced results matching real-world growth rates and nearly eliminated the Solow residual, he also clarified the importance of energy to the economy.

What he found was that standard neoclassical economic theory badly underestimates the role of energy in fuelling economic growth. In fact, it's wrong by a factor of ten. In the case of West Germany, a 1 percent rise in energy input resulted in a 0.5 percent rise in GDP, not the 0.05 percent Solow's theory predicts. In the United States, an increase in energy boosted growth even more dramatically. In Kümmel's words, "the economic weight of energy remains high above its cost share,"[6] and we pay for energy only about a tenth of its actual economic worth.

This cheap energy circulating in the economy is the equivalent of work done free by "energy slaves." Kümmel divides the amount of energy consumed in the economy by the number of calories a human needs to live and work and comes to the conclusion that the average American has ninety energy slaves. Who would do all that work if the energy slaves disappeared?

More recently, Robert Ayres and other researchers have correlated economic growth with increases in thermodynamic efficiency—the ratio between the work a unit of energy can potentially do and the work it actually does. This research shows that Kümmel's calculations likely underestimate energy's role in growth. Ayres notes that efficiency has improved over time. A horse, for example, turns feed into work at a thermodynamic efficiency of about 4 percent. Since 1910, the efficiency of electric-power generation and distribution has gone up from about 4 percent to nearly 35 percent. According to Ayres, our economies expand not so much because our energy inputs grow as because our efficiency in using these inputs rises. With this refined methodology, Ayres almost completely eliminates the Solow residual.[7]

If we're to have continued growth without steady increases in fossil fuel consumption, Ayres notes, we will have to find ways to reduce the amount of fossil fuel energy input per unit of work. Energy conservation and energy efficiency are key to economic and environmental sustainability. But the more salient conclusion of this recent research is that our economies are deeply dependent on massive inputs of cheap fossil energy.

Something so vital to our civilization probably won't stay cheap for long. The price spikes for oil and gas in 2008 suggest that energy markets are heading towards a new and dearer equilibrium. Not only are people beginning to realize how important oil is, traders and investors are starting to fear there won't be enough of it to go around.

Even oil industry insiders are beginning to admit this hard truth. In 2007, the U.S. National Petroleum Council, made up of top oil and gas industry executives and chaired by Lee Raymond, the hard-nosed former CEO of ExxonMobil, called for energy and petroleum conservation, saying that "expansion of all economic energy sources will be required to meet demand reliably" till 2030.[8] The Council acknowledged that this requirement imposes severe "compound challenges." Put simply, it will take every energy source we can tap to sustain even the possibility of business as usual through the next quarter century.

Whether or not we face an imminent crisis of petroleum supply (a topic debated in the essays that follow), a few things seem intuitively obvious. First, fossil fuels in general and oil in particular are a physically finite resource. This resource is a precious inheritance of solar energy that warmed the planet millions of years ago. There's only so much of it in Earth's crust, and when the stuff in the ground is gone, it's gone for good. Of course, the question is more complicated than this, because we can't know for certain how much fossil fuel is in the ground or even how much of the resource that we know is there will be extracted: technology makes available sources of oil, gas and coal that were previously unreachable or too dispersed, while rising prices make it profitable to tap reserves that previously would have produced a financial loss. But despite these uncertainties, it's clear that we can't mine coal or pump oil and gas indefinitely.

The proof that conventional oil is a finite resource is readily at hand. The United States was once the biggest oil producer and exporter in the world. Just as OPEC (Organization of Petroleum Exporting Countries) today sets production quotas for its members in order to keep prices high, for decades Texas kept the taps partially closed in order to avoid glutting the market. Now the U.S. is the biggest oil importer on the planet. Not long ago, the United Kingdom was also a key exporter; now the British have to import oil, and it's expected that by 2020 the North Sea will be producing only one-fifth of what it was producing at peak in the late 1990s.[9] Today, many of the world's most important oil-producing countries have passed their peak conventional oil output, and their production is dropping by 3 to 4 percent every year or more (U.K. production is falling 8 percent a year). They're not out of oil. On the contrary, they're still pumping and most are still exporting. It's just that they will probably never achieve the output levels they saw at their peak. Once an oil-producing region passes peak production, output generally slides downhill, no matter what is done.

As one conventional oil field goes into decline, another has to be brought online, or production (possibly nonconventional production, like that from Alberta's tar sands) has to be ramped up elsewhere to make up the loss. But conventional discovery is lagging: we now consume between six and ten barrels of oil for each one we discover. Global discovery peaked in 1964 and has been falling ever since (interrupted by a brief rise in the 1970s when the North Sea and North Slope fields were found). Also, today's discoveries are generally smaller and more difficult to exploit, and the oil from them is generally more expensive to refine than was the case with the giant (and now very mature) fields of Saudi Arabia and Texas. It remains to be seen how many years will pass between the global peak of conventional oil discovery and the peak of conventional oil production—in oil-producing regions around the world, the lag has ranged from fifteen to forty years and has generally dropped as extraction technologies have improved. But, once again, it seems intuitively obvious that there will be a peak sometime. What's less obvious is when it will happen and whether it matters.

Most experts agree that the "when" question will be answered only in hindsight (although many geologists who've been sounding the alarm about peak oil estimate that the global peak will arrive sometime between 2006 and 2010). Basic economics tells us that as we reach peak global production, oil prices will rise, perhaps sharply. These higher prices will in turn depress oil demand—and, ultimately, pressure on oil supply. The result might be an extended period during which global oil production bounces off a relatively firm ceiling, as price spikes are followed by global recessions, and subsequent economic recoveries are followed by further oil price spikes. This may be the situation we find ourselves in now: the record oil prices of early to mid-2008 gave an extra push to a downturn in the U.S. economy (mainly caused by a collapsing housing market), and then falling oil demand allowed prices to plummet later in 2008.

So the peak may already be here, except for a while it may look more like an extended plateau. Global conventional oil production has stayed constant at about eighty-six million barrels per day since 2005. Significant new production—about three million barrels a day—will come online in 2009, mainly from Saudi Arabia, the Gulf of Mexico and Kazakhstan. But whether this production boosts overall global output beyond eighty-six million barrels a day depends on decline rates in mature fields elsewhere—a factor surrounded by uncertainty. Experts have been surprised by the speed of decline in the last few years in Mexico and the apparent difficulty Russia is having maintaining its high production. Nonetheless, the extra production in 2009 combined with weaker demand could easily depress oil prices further. But the view from 2010 and beyond is not so rosy—only modest new production is slated to come online in 2010 and 2011. If the world economy starts to recover, oil prices will likely rise once again.

Will it matter? Optimists like Mark Jaccard, who has a chapter in this book, say probably not. Any deficit between conventional oil supply and demand, they argue, will be met as higher prices stimulate oil conservation by consumers, increased drilling by oil and gas companies, high-tech extraction of more oil and gas from existing mature fields, and a shift to other sources of liquid fuels, including non-conventional heavy oils like Alberta's tar sands. Other experts would argue that the evidence to support this optimism is mixed. On one hand, there is little long-term correlation between oil price and oil discovery: the biggest conventional oil discoveries have generally occurred in times of relatively low oil prices, and many periods of high oil prices, like the period following the price doubling in 1979–80, have stimulated remarkably little discovery. On the other hand, new petroleum extraction technologies can dramatically boost reserves. To take just one example, recent breakthroughs in horizontal drilling and the use of high-pressure water to shatter rock have allowed drillers—in the last three years—to start extracting enormous amounts of natural gas from what previously were unexploitable oil shales in North America, reversing what had otherwise been a steady decline in the continent's gas production.

On balance, though, the longer-term global challenge looks truly formidable, especially when it comes to oil. Even as the planet's largest conventional oil fields, from Saudi Arabia and Kuwait to China and Mexico, peak and go into decline, global oil demand will keep soaring and is expected to rise 50 to 60 percent by 2030.[10] Where is all this oil going to come from?

It's nearly impossible to exaggerate oil's importance to our economies. We depend on oil far more than the Lydians depended on gold, and the history of the second half of the twentieth century shows just how vulnerable we are to disruptions in oil supply. When oil prices quadrupled in 1973, the world economy went into reverse; in the United States, economic output fell by 6 percent and unemployment doubled.[11] When the Iranian Revolution in 1979 caused oil prices to double, the globe tipped into a recession. The same thing happened in 1990, when Iraq invaded Kuwait and oil prices briefly soared. If, for whatever reason, oil suddenly becomes scarce and prices rise sharply, the gears of our economies begin to seize up.

High prices mean a transfer of wealth from oil-consuming to oil-producing countries. They mean lower personal income in the consumer countries, because higher household spending on energy means there's less to spend on other goods and services. Higher oil prices also mean inflation, unemployment, reduced capital investment, diminishing tax revenues and growing budget deficits. A report from the International Energy Agency estimates that a ten-dollar increase in the price of a barrel of oil shaves 0.4 percent from a rich nation's GDP and as much as 3 percent from a sub-Saharan African GDP.[12] According to a report for the U.S. Department of Energy, for every doubling in the price of oil, the American economy loses a full percentage point of GDP.[13]

Other factors exacerbate the problem of dwindling conventional oil. Even if the oil is accessible in the ground, shortages of drilling rigs, rig workers, deep-sea welders, oil tankers and refining capacity are limiting our ability to get the oil on the market fast enough to meet global demand. Global tanker capacity and refining capacity both peaked in 1981, as did the number of oil rigs. Some $2.4 trillion worth of infrastructure will have to be built within the next decade to meet demand and avoid crisis—nearly three times what was spent in the previous decade.[14]

Then there are geopolitical factors. Russia's proximity to the Baku-Ceyhan pipeline, Israel's threats to bomb Iran, Iran's threats to close off the Straits of Hormuz, insurgent attacks against offshore rigs in Nigeria, and America's worsening relations with Venezuela all keep global oil markets on a knife's edge of uncertainty. At any moment, we're probably one major geopolitical crisis in an oil-producing region away from a doubling of crude oil prices, virtually overnight.

Whether or not 2008's price spike signalled the arrival of conventional oil's peak, it sent ripples through the North American economy. The big three North American car manufacturers reported one record loss after another. Some of their factories went silent as autoworkers were sent home by the tens of thousands. Showrooms full of sport-utility vehicles and muscle cars—where people were looking for personal expression and (upward) mobility—suddenly emptied of customers. The airlines too, once symbols of twentieth-century excitement and glamour, cancelled flights, laid off employees and mothballed jets. Transit ridership in urban centres soared, while house prices fell fastest in the suburbs, because many suburbanites decided they couldn't cope with the cost of multi-hour commutes. Municipalities tried to cover higher fuel prices by cutting back on street and park maintenance, community centres' hours of operation, and emergency services. Meals on wheels programs were cancelled.

In these countless large and small ways, business as usual seems to be ending.

—

But even if we can afford to pay for fossil fuels, can we afford to burn them? The greatest threat to our future may be not that our fossil fuel economy will disappear—but that it will endure.

That's because at the same time that we've been enjoying the benefits of a modern industrial economy, we've been releasing huge amounts of carbon dioxide into the atmosphere, mainly by burning fossil fuels. In the mid-nineteenth century, the atmosphere's carbon dioxide concentration stood at about 270 parts per million. Now it's at 387 parts per million—higher than at any point in the past 650,000 years and rising at more than about 2 parts per million annually. Despite all our eco-shopping and government commitments to do the green thing, we're adding about thirty billion tonnes of carbon dioxide to the atmosphere annually, equivalent to the mass of 280,000 *Nimitz*-class nuclear aircraft carriers. (The amount of carbon dioxide we generate is much larger than the amount of carbon-based fuel we consume. In the process of burning the fuel, its carbon combines with oxygen—that is, one carbon molecule joins with two oxygen molecules, creating a total mass almost four times heavier than the original mass of carbon. So the carbon dioxide an airplane emits while it flies is actually heavier than the fuel it carries.)

Scientists have known for more than one hundred years that adding a lot of carbon dioxide to the atmosphere would raise the planet's temperature. They've been observing the predicted changes for decades and warning policy-makers about the implications of humankind's behaviour. But the knowledge and warnings haven't really made any difference. We've continued to dump staggering amounts of climate-changing carbon dioxide into the sky—not just by burning fossil fuels but also by cutting down forests, making cement and engaging in countless other common industrial and consumer activities. Meanwhile, evidence is mounting that the soils, forests and oceans that have, in the past, absorbed about half the carbon humans have released are now starting to saturate—or even, as the climate warms, beginning to flip from being net sinks for carbon to net sources. As James Lovelock explains it, "What we are now doing is uncannily like the series of foolish actions that led to the Chernobyl nuclear reactor accident. There the engineers turned up the heat after they had disabled the safety systems, and it should have been no surprise that the reactor ran into rapid overheating and caught fire."[15]

Until recently, climate scientists generally agreed that 450 parts per million was potentially a critical threshold. Beyond that point, they argued, the biosphere itself could start to release massive amounts of carbon on its own, as melting permafrost emits methane, forests die from disease and burn and the ocean's critical ability to absorb carbon drops sharply (in part because of lower productivity of carbon-eating phytoplankton). This would be the beginning of a slide to irreversible and perhaps catastrophic climate change. But in the last two years, a consensus has begun to emerge that the safe threshold is likely far below 450 parts per million. James Hansen, the lead climate scientist at NASA (National Aeronautics and Space Administration) and a key voice in the fight for better climate policies for thirty years, has recently argued that the world needs to return to 350 parts per million. There's only one problem: that level was passed two decades ago.

So it's quite possible that even the activists at the forefront of the campaign to keep the planet from overheating haven't been ambitious enough. The rigour of climate science makes its findings and predictions inherently conservative. But real-world observations—in the Arctic and elsewhere—are now often outstripping the results of scientists' best forecasting models. Global warming's expected changes are arriving years—and sometimes decades—ahead of schedule.

According to the Intergovernmental Panel on Climate Change (IPCC) scenario for future emissions that best represents "business as usual"—meaning an extrapolation, decades into the future, of the current compounding growth of greenhouse gas emissions—we can expect average global surface temperature to rise somewhere between two and four degrees Celsius this century, with a most likely increase of around three degrees. This projection includes only a few relatively simple feedback effects—whereby greenhouse warming produces changes that cause more warming—because scientists' knowledge of many feedbacks is still rudimentary. So actual warming could be considerably worse.

To put the projected amount of warming in context, the total increase in Earth's average surface temperature since the coldest period of the last ice age fifteen thousand years ago has been about five degrees Celsius. The last time Earth's average temperature was two degrees above the 1900 level was 130,000 years ago, during the previous episode of warmth between glaciations, and sea levels then were four to six metres higher than they are today. The last time the temperature was three degrees above the 1900 level was about thirty million years ago, and sea levels were twenty to thirty metres higher than today.

While sea-level rise of this magnitude will likely take centuries, even within this century shifting weather patterns could
turn the world's breadbaskets into dust bowls; melting glaciers
and ice sheets in Greenland, Antarctica and elsewhere could
raise seas enough to displace hundreds of millions of coastal
dwellers; heat waves could kill hundreds of thousands of people
in the world's cities; loss of meltwater as glaciers disappear could
dry up rivers on which billions of people depend for fresh water;
and higher temperatures and extreme weather could lead to
waves of extinctions.

Such changes would impose a staggering cost on humanity,
perhaps as large—suggests Nicholas Stern in his 2007 report,
The Economics of Climate Change—as the greatest calamities of
the twentieth century, the two world wars and the Great
Depression. The insurance industry is already deeply alarmed.
Losses from natural disasters have been soaring over the past
few decades; in the 1990s they were fifteen times what they
were in the 1960s.[16] The crippling cost and difficulty of rebuild-
ing a small city like New Orleans in the wake of a glancing
blow from Hurricane Katrina gives some sense of the vulnerability of the complex networks we've built for ourselves.
Katrina nearly overwhelmed several insurance companies, just
as the storm surge washed over the levees. Only one or two
Category 4 hurricanes slamming into the heavily populated
coastline of a rich country could lead to claims in excess of the
reserves the insurance industry sets aside to cover these losses,
triggering a concatenating series of financial crises.

Climate change is also emerging as a threat to national
and international security. Indeed, it may well represent a
security challenge just as dangerous as, and ultimately more

intractable than, the arms race between the U.S. and Soviet Union during the Cold War or the proliferation of nuclear weapons to rogue states today.

Climate change will help produce exactly the kind of military challenges that today's conventional forces don't cope with well—violence in the form of insurgencies, guerrilla attacks, gang warfare, and terrorism that's diffuse, chronic and subnational. In already vulnerable poor countries, climate change will increase the frustrations and anger of hundreds of millions of people by weakening rural economies, boosting unemployment and dislocating people's lives. Especially in Africa, but also in some parts of Asia and Latin America, it will undermine already frail governments—and make challenges from violent groups more likely—by reducing revenues, overwhelming bureaucracies with problems and revealing how incapable these governments are of helping their citizens.

A recent report from the Center for Strategic and International Studies in Washington, DC, offers scenarios under conditions of "expected," "severe" and "catastrophic" climate change. The "expected" scenario is, as the report puts it, the "best we can hope for" and at the same time "the least we should prepare for." It's based on the low end of the IPCC projections of just over one degree Celsius warming by 2040. Yet it's still "a scenario in which people and nations are threatened by massive food and water shortages, devastating natural disasters, and deadly disease outbreaks."[17]

The "severe" scenario takes into account some of the feedbacks left out of the IPCC projections and so assumes warming of 2.6 degrees Celsius by 2040. It also reflects the principle that "nonlinear climate change will produce nonlinear political events." In addition to more frequent and more violent storms and the abandonment of major cities, this scenario includes the breakup of the American federal political system and a much higher likelihood of severe tension between the U.S. and Canada. The "catastrophic" scenario, which assumes 5.6 degrees Celsius warming by 2100, predicts "almost inconceivable challenges as human society struggled to adapt."

So the phrase "climate change" doesn't begin to capture what's in store for our children and grandchildren. It is not just that the weather will change. Our human civilization, which for its brief existence over a few thousand years has flourished in a benign climate, will suddenly find itself in a world that's radically different.

In response to this profound, even existential challenge, it's clear that we have to do something—and likely something drastic. In coming decades, we're going to have to cut global carbon emissions deeply. The main question is: how much? Most experts are now calling for cuts of at least 80 percent by 2050.

In the end, the best-case scenario and the worst-case scenario turn out to be the same thing: business as usual. The best we can hope for is that we don't run out of cheap oil, and the worst we have to fear is that we will continue to burn fossil fuels, including oil, as we've burned them in the past.

Nothing illustrates the convergence of the twin, coupled crises of worsening scarcity of cheap oil and climate change better than the diminutive Tata Nano, the little Indian car that promises to do for the modernizing Indian economy and fledgling middle class what the Ford Model T did for the common man in the United States about a century ago. It will turn a nation of more than a billion people into an automotive society. At a price of only about twenty-five hundred dollars, the Nano will be within range of about two hundred million Indian drivers, almost all of whom get around today on mopeds, bicycles and rickshaws. By 2015, Indians are expected to buy three million new cars each year. As Jeff Rubin points out in his essay here, Indians aren't the only ones lining up to live the dream of happy motoring. By 2015, China's car market is expected to overtake that in the United States, with sales doubling to seventeen million new cars a year. And other developing economies are on the same path.

All those new Nanos are going to need gasoline and diesel to keep them going, at the very moment that global production of conventional oil is likely to have plateaued. And all of them are going to emit carbon dioxide, just when we need to make massive cuts in global emissions. Also, making these cars and building the roads and bridges to accommodate them will require vast amounts of energy—the consumption of which will emit vast amounts of carbon dioxide. The process of making a tonne of cement releases about a tonne of carbon dioxide. China led the world in cement production in 2007, churning out 1.4 billion tonnes of the stuff (followed by India and Russia). Filling up the tanks of the world's cars is not the only thing that pushes us closer to the limits of our geological resources and well past the limits of the planet's ability to absorb carbon dioxide. So does the burden of manufacturing them, servicing them and building places for them to go.

What makes the twin crises of worsening scarcity of cheap oil and climate change particularly intractable is that we must solve both at once, because each makes the other harder.

The most obvious solution to our conventional oil problem (turning to carbon-intensive sources of liquid fuels, like diesel made from coal or oil made from Alberta's tar sands) will make climate change a lot worse, while the most important first step to deal with climate change (restricting the use of the most carbon-intensive sources of energy) will make the economic impact of scarcer conventional oil a lot worse.

The prospect of running out of cheap oil is a source of hope for some. In only a few weeks, the spectacular rise in the price of a barrel of oil in 2008 resulted in lifestyle changes environmentalists have been advocating for decades. Commuters started taking public transit in record numbers, businesses started using teleconferencing rather than air travel, fuel-efficient cars started flying off the dealers' lots. But high oil prices also made the use of carbon-intensive energy sources look far more appealing. Prime Minister Harper billed Canada an "energy superpower" because of the potential of Alberta's tar sands, ignoring the fact that total carbon emissions of gasoline derived from the tar sands are 30 to 70 percent higher than those from gasoline derived from conventional oil.[18] And liquid fuel made from coal also started to look much more profitable, even though this fuel is far worse, in terms of carbon intensity, than tar sands oil—with total emissions 85 percent higher than conventional gasoline.[19] Worsening scarcity of cheap oil may make us a little greener on one hand, but it will certainly make our economies much dirtier, if the market is left to decide on its own what we'll use to replace that oil. As David Keith argues in the following pages, there's enough carbon energy beneath our feet to produce climate change that would make the planet uninhabitable.

This highlights the kernel of the climate–energy problem. It's not oil or gas or coal in particular that threatens us. It's carbon. The converging problems that loom in the coming decades are in fact the same problem—a carbon problem. It's the carbon in the fossil fuel we have become so dependent on that threatens us.

And it's not the price of any particular energy source that will push us in the right direction, but the price of the carbon it contains. Expensive oil will push us towards alternative fuels, as Mark Jaccard argues. But these new fuels are likely to be worse for the climate than the oil they replace. Expensive carbon, however, will trigger a migration towards something cleaner. We must meet our huge (and growing) energy demand with a new source of energy that's both energetically cheap and far lower in carbon intensity. Find energy without carbon and the riddle is solved. This is what we mean by a *carbon shift*.

But to achieve a carbon shift we need a better understanding of the dimensions and the scope of the carbon problem. It's completely rational for any of us—the neighbour driving a sports-utility vehicle or the government of Alberta ramping up tar sands production—to pursue our own interests under the rules that prevail in our economy and society. Unfortunately, our misunderstanding of the problem has obscured our real interests, while market rules that make fossil energy too cheap and give us the wrong economic incentives have skewed our behaviour in a direction that hurts future generations. The solution will be to better our understanding, redefine our interests and change the economic rules.

This book is an effort to help bring about such transformations. It's meant to put readers in direct contact with experts on our energy and climate challenges. High prices at the pumps may provoke a few news stories about oil supplies or hybrid car sales, but only rarely does a geologist such as David Hughes, probably Canada's leading authority on the question of our fossil fuel inheritance, have his say in a public forum. Here, in this book, he gives us a rigorous sense of how much oil is left in the ground and how hard it will be to get it out.

When issues are enormously complex, as are the issues covered in this book, experts inevitably disagree. Hughes argues that peaking oil production is the biggest problem facing our economies and ways of life. But David Keith, among the world's most respected authorities on the intersection of climate and energy problems, makes the case that the climate problem is a far more urgent crisis. And Mark Jaccard, energy economist and former IPCC member, takes a different view, arguing that economics rather than geology will define our energy future. As long as we can provide markets with the right price signals, Jaccard contends, we can count on them to keep our complex systems running. Economist Jeff Rubin, a globally recognized voice on the threat of oil depletion, argues on the other hand that shifting patterns of oil demand are challenging our long-held economic assumptions. Rich countries are no longer driving petroleum markets, but are just along for the ride as rapidly industrializing poor countries, like China and India, push their accelerators to the floor.

Canada has its foot to the floor too—in the tar sands. Oil companies and investors dream that Alberta will be the next Saudi Arabia. In his chapter, journalist William Marsden lays out the dream's costs and compares them with its likely benefits. Journalist Jeffrey Simpson then holds a mirror up to all Canadians, asking whether we really want the decisive action we say we want, or whether we'd rather just continue to defer costs of action into the future—in the hope that someone else will foot the bill.

None of the contributors would claim to have settled any of the complex matters they discuss in the few short pages at their disposal here. But the point of this book is not to solve these problems but to invite readers to see them through the eyes of those who think about them rigorously. Imagine finding yourself sitting on a train beside a leading economist or scientist or journalist who spends day after day studying the issues that will define the future of so much you and your family hold dear. You'd want to use the short time with this person to learn exactly what he or she thought. You might want to seek out other opinions later or perhaps do your own research. But one way or the other, direct access to that expert opinion would likely change your outlook and could even spur you to action. Such change is what *Carbon Shift* is about.

These contributors answer many questions—but they raise more. And that's because the carbon problem is particularly intractable. Like the gold in the river Pactolus, carbon in its various forms is both precious and deeply harmful. Until we understand this dilemma—and crack it open—we run the risk of solving only half the crisis, and in this case half is not nearly good enough.

Thomas Homer-Dixon and Nick Garrison

David Keith
DANGEROUS ABUNDANCE

David Keith has worked near the interface between climate science, energy technology and public policy for twenty years. His work in technology and policy assessment has centered on the capture and storage of CO_2, the economics and climatic impacts of large-scale wind power, the prospects for hydrogen fuel, and the technology and implications of global climate engineering. Keith has built an infrared spectrometer for NASA's ER-2 and developed new methods to increase the safety of stored CO_2.

Keith has held the Canada Research Chair in Energy and the Environment and served as Professor in the Department of Chemical and Petroleum Engineering and Department of Economics at the University of Calgary since 2004 and remains an Adjunct Professor in the Department of Engineering and Public Policy at Carnegie Mellon. He was named environmental scientist of the year by *Canadian Geographic* in 2006.

Keith has served on numerous high-profile advisory panels such as the UK Royal Society's geoengineering study, the IPCC, and Canadian "blue ribbon" panels and boards. David has addressed technical audiences with articles in *Science and Nature*; he has consulted for national governments, global industry leaders and international environmental groups; and has reached the public through venues such as the BBC, NPR, CNN and the editorial page of the *New York Times*.

We are simultaneously threatened by both the scarcity and the abundance of fossil fuels. The scarcity of conventional oil poses a serious threat to economic and geopolitical stability, a threat that is intensified by the hugely unequal geographic distribution of the remaining easy oil. Yet, while oil grows scarcer, the very abundance of fossil fuel resources poses a threat that is at least equally serious: climate change—the greatest global environmental threat of our age—is rooted in the extraordinary abundance of fossil resources and the growing ease with which our technology can exploit them.

The scarcity of easy oil tempts many observers to assume that humanity faces an immediate and far-reaching crisis of energy supply, and it is not hard to see why. It is difficult to overstate the importance of energy in modern society. Coal, gas and oil enabled the world's industrial transformation. Many of the technologies and institutions of the modern world first emerged during the eighteenth century's "enlightenment," but it was access to cheap energy that played the central role in accelerating the Industrial Revolution a century later. Access to abundant fossil fuels—first coal, then oil and now gas—has driven the growth in human population and led to the mobility, high-speed communications and widespread—though grossly unequal—material abundance that are the hallmarks of our age.

Yes, oil is crucially important. But let's not confuse oil with energy. Oil may be scarce, but energy is not. A great deal of energy is stored in fossil fuel reserves which, in their abundance, pose a threat at least as serious as the prospect of running out of oil. There are enough fossil fuels beneath our feet to push atmospheric concentrations of carbon dioxide to well over ten times their pre-industrial levels. We may well have enough fossil fuel within the growing reach of our extraction technologies to nudge our planet's climate towards that found on Venus. Not Venus the goddess of love, but Venus the planet, where the atmosphere is 95 percent carbon dioxide and surface temperatures are hot enough to melt lead. So while it is possible to make a case that oil scarcity poses a threat to our civilization, I argue that fossil-energy abundance is where the more urgent threat lies.

In this essay, I explore the interplay between oil scarcity and climate change, shaping the argument around two alternative scenarios. In the first, we will assume that an oil peak is imminent and explore our options while ignoring concerns about climate change. In the second scenario we will assume the converse: a world with abundant oil in which carbon emissions must be swiftly eliminated to minimize the risk of dangerous climate change. The two scenarios bound the likely path to the future; enabling us to see the some of the more onerous constraints and through them to imagine future opportunities.

Scenario One: Tight Oil, Loose Carbon

Suppose that we lived in a world where oil production declined sharply yet there were no concerns about carbon and climate. How might we reshape the energy system to answer this threat? To put some meat on the story, let's assume that the production of conventional crude oil peaks today and begins to decline, at first slowly and then at an accelerating rate, so that production is a third of today's roughly eighty-five million barrels per day by 2050.

This is a strong "peak oil" scenario in which the decline in production is a little faster than the rate at which production increased over the last forty years. My view is that such a scenario is quite unlikely. As I will describe below, there are good reasons to expect that production of conventional oil would decline more slowly, unless demand is intentionally restrained. However, peak-oil advocates make some strong arguments, so it's worth considering such a scenario. In any case, my purpose here is to examine what we could do to manage a sharp oil peak—ignoring the climate constraint for now—independent of the probability that such a peak will actually occur.

On the climate side, one could assume—for the purpose of the scenario—that some cheap and clean method is found for removing carbon from the atmosphere. For argument's sake, suppose someone discovers a cheap method of accelerating the global weathering cycle that drives the permanent removal of carbon dioxide from the biosphere and so eliminates the climate threat. The mechanism is unimportant. All we have to do is envision a hypothetical world in which we don't have to worry about the climate. Could we manage an oil peak under this scenario?

The short answer is yes.

At this point, let us ignore the political economy of scarce oil and focus instead on technology and costs. The technical solutions to oil scarcity depend on transforming other energy sources into petroleum substitutes, and so all these options rest on one central fact: *there is no shortage of energy.* The age of conventional oil pumped from the ground will surely end, either as a gradual peaking and slow decline over a century, or the more dramatic decline I am considering here. Come when it may, the end of easy oil will not signal a shortage of fossil fuels or of energy. We have sufficient resources of coal, gas and unconventional oil to power our fossil-driven civilization at twice our current burn rate for more than two centuries.

How much fossil energy does the earth hold? In considering the question, one naturally imagines a tank with a fixed capacity, but this is a fundamental misconception that frustrates any attempt to make sense of the acrimonious debate about the future of fossil fuels. Imagine instead a line of fuel tanks stretching into an uncertain distance, each with a definite capacity and cost of extraction. Standing at the head of the line we see the conventional oil reserves that are producing today. Looking farther down the line, we cannot read the label with each tank's precise capacity and cost, but we can see that there are huge resources such as coal at depths too deep to mine with today's technology and oil that is in such small or hard-to-reach pockets that it is now uneconomic to extract. But it is there.

Economic geologists break the line of barrels into reserves and resources. Reserves are the amount of fuel that can be extracted at current prices using current technologies. If prices go up or if technology drives down the cost of access, then reserves expand without anyone discovering a drop of new oil. Reserve-to-production (R/P) ratios are the number of years that a reserve would last at the current rate of production. In contrast to reserves, resources are the estimates of the amount of fuel in the ground both discovered and undiscovered.

The gradual conversion of dimly seen resources to producing reserves that shapes the future of fossil fuels is best seen in the United States, the country where the oil age began, and the country with the largest cumulative production of oil. A century and a half after production first began, and almost four decades after oil production peaked in 1970, the R/P ratio in the United States now stands at twelve years. Is the U.S. twelve years away from the end of domestic production? Not at all. R/P ratios have been similarly low for decades, and yet total petroleum production has been remarkably constant, falling only about 32 percent in the last twenty years. U.S. production is declining, but the rate of decline is far slower than that suggested by a glance at current reserves; we should expect similar dynamics for the world's oil production as resources are gradually converted into reserves.

So how big are these resources? By definition, we don't know exactly. Estimates are uncertain and contested. However, they are all immense in comparison with current consumption. Common estimates of the combined amount of reserves and resources of oil and gas range from three to six trillion barrels of oil equivalent, while for coal the total is between twenty and fifty. These estimates include unconventional resources, such as tight gas and bitumen in oil sands, but ignore gas hydrates that contain an energy equivalent of about 130 trillion barrels of oil. These resources estimates dwarf current consumption rates of oil, gas and coal, which now stand at twenty-seven, nineteen and twenty-two expressed in billions (one thousandth of a trillion) of barrels of oil equivalent per year.

Put another way, the energy content of fossil fuel resources (ignoring gas hydrates) is equivalent to between four hundred and eight hundred years of global energy use at current rates of consumption. So it is a little early to say we are running out.

How can these resource estimates be squared with the evidence of resource scarcity given by advocates of "peak oil"? In general terms the answer is that those who argue for rapid exhaustion of oil, gas and coal do so primarily from curve-fitting analysis which ignores much of what we know about economics, geology and technological change.[1] The strongest arguments generally apply to oil where a good case can be made that Mideast oil reserves are systematically exaggerated. Likewise, the "Hubbert's Peak" resource production curves do a good job explaining the historical production of U.S. oil for which scarcity is clearly the driving force.* However, the application of similar curves to U.S. coal production, for example, is virtually meaningless since they involve implausible extrapolations from early points on a curve to its endpoint; and, since they implicitly depend on assumptions that coal production rates have been driven by resource scarcity while the history of U.S. coal prices (which have declined) contradicts this assumption, which in turn makes a mockery of the underlying rationale behind the curve fits.[2]

---

* Editor's note: Mark Jaccard provides an example of a Hubbert curve on page 111.

There are multiple technological pathways to providing an increasing energy supply with a decreasing environmental footprint (ignoring carbon) at a cost that is a fraction of what we pay for health care. I will describe two of these important pathways: first, synthetic fuels and then, electrification, treating each in some detail to give a sense of the opportunities, costs and technical barriers to their large-scale implementation. Many other options exist, including biofuels, hydrogen and radically accelerated increases in efficiency, but I will ignore them here since they have been widely discussed in energy debates and may be more familiar to most readers.

Synthetic Fuels

Synthetic fuels are chemically similar to existing fuels such as gasoline or diesel but they are derived from a source other than crude oil.

Coal is the most common feedstock for synfuel production. A typical coal-to-liquids process starts by gasifying coal to produce a so-called syngas (synthesis gas) composed of hydrogen and carbon monoxide along with carbon dioxide and a mess of minor constituents. The hydrogen and carbon monoxide are then combined in a catalyst bed, composed of metals such as iron or cobalt, to make hydrocarbons such as octane, the central component of gasoline.

This is old technology. While gasification is sometimes portrayed as a high-tech challenge for the energy industry, it has roots extending back more than two centuries, when deforestation forced European iron makers to replace charcoal with coke made from coal.

Indeed, "natural gas" got its name because when it entered the market, it was replacing a syngas called "town gas" that was made from coal by using a close relative of today's gasification process and then distributed through pipeline networks in many cities in North America and Europe. In these places, natural gas did not overtake town gas production until after World War II.

Synthetic liquid fuels also have a deep history. During World War II, Germany had little oil but abundant coal so it built huge coal-to-liquids facilities to enable fuel-intensive warfare with limited oil supply. Synfuel production supplied more than 90 percent of Germany's aviation gasoline and half its total liquid fuels.

Coal-to-liquids turns a low-cost feedstock into a high-value product with the assistance of a large helping of expensive infrastructure. Coal costs five to ten dollars for the energy equivalent of a barrel of oil, and current synfuel technologies require about two energy units of coal to make one energy unit of gasoline, so the cost of coal required to make a barrel of energy should be less than twenty dollars. When interest on capital and production costs are included, the long-run cost of refined synfuel products such as gasoline or diesel is in the order of one hundred dollars per barrel of oil equivalent.

While the process technologies are very different, coal-to-liquids and oil sands share a similar niche in the energy system. Both turn a low-cost energy feedstock into a high-value product: liquid fuels for transportation. Both require large inputs of capital and have high operating costs, and in both cases the cost of the raw material—coal or bitumen-rich sands—is a small component of the overall cost of producing the fuel. Finally, both take a high-carbon fuel as input and produce a lower-carbon fuel and relatively large amounts of carbon dioxide emissions, a topic we will return to when we consider climate-constrained futures.

Could synfuel plants be built quickly enough to fuel the growing transportation demand in a peak-oil world? The resources and technology are at hand, but the answer depends strongly on one's judgment about the investment climate for this kind of large energy infrastructure. For this reason, I will attack the question from three alternative viewpoints.

One could argue for the negative by noting that governments of the major industrial nations invested several tens of billions of dollars into synfuel development following the 1973 oil shock, yet there has been very little production to date. While this argument carries some weight, the lack of investment probably has had more to do with the decline in oil prices at the time this research would have been brought to market in the early 1980s than it has had to do with technological failures. Business has been sensibly shy about investing in coal-to-liquids, a high-capital-cost technology where the payoff depends so strongly on oil prices that lie outside the developer's control.

An argument for the positive is based on capacity and need. Current estimates of the investment cost for coal-to-liquids facilities are about one hundred thousand dollars for each barrel per day of capacity, so a million barrels per day of capacity could be brought online each year at an investment cost of one hundred billion dollars. The cost estimate above is inflated to reflect the high construction costs of the current energy boom, yet even if one assumes that capital costs were to double again because of the rapidity of construction, the

United States could build enough coal-to-liquids capacity to eliminate the need for oil imports within twenty years at an annual investment cost of about 1 percent of gross domestic product (GDP). If oil security was truly seen as an existential threat to the welfare of developed nations, there seems little doubt that such investments would be made. Indeed, China is now making huge investments in coal-to-liquids plants because of the government's fear of oil scarcity.

Finally, could we build large-scale coal-to-liquids facilities without wrecking the environment? Putting aside carbon dioxide emissions for the next scenario, I expect it would be possible to produce fuel from coal with less environmental damage than current oil and gas production, because while it's locally destructive, coal mining and processing disturbs less land per unit of energy than does modern oil and gas production. But my judgment hangs on an admittedly idiosyncratic view that sees land use as the central long-term driver of environmental damage. All the same, there is no doubt whatsoever that if coal-to-liquids plants were built quickly as a chaotic response to a real or perceived crisis, the environmental damage could be horrific.

Electrification

At current Canadian prices, it is about five times less expensive to provide a unit of energy to the wheels of a car using electricity than using gasoline, when one accounts for the higher efficiency of electric motors. Why don't electric cars dominate the market? Because the capital cost of batteries overwhelms this operating-cost advantage.

Transportation is the only important sector of industrial economies that has not been transformed by electrification. Electricity storage is the weak link that breaks the chain from power plant to wheels, limiting electricity to niches such as electric trains.

In the twelve decades since Nikola Tesla's invention of alternating currents opened the electric era, we have devoted an ever-increasing share of our primary (raw) energy inputs to electricity, so that now more than one third of all energy used worldwide is converted to electricity before being put to work. Electricity is an attractive energy carrier because at point of application the devices that transform it into useful energies, such as motive power, heat, light or computation, are generally efficient, compact, reliable, non-polluting and inexpensive. And, by decoupling primary energy inputs from energy use, the electric grid allows us to blend energy inputs such as coal, wind and nuclear heat into a uniform fuel. Your toaster or stereo can't tell the difference when you plug it in. Just as we can turn coal into oil, we can turn just about any energy resource into electricity.

The efficiency of modern electric systems is a marvel: the overall efficiency from power plant to wall plug exceeds 90 percent in most rich countries. That is, only 10 percent of the power leaving a big power plant is lost in the wires between the plant and an average consumer. Many small electric motors exceed 70 percent efficiency and large industrial motors have efficiencies of more than 95 percent; and finally, the round-trip (energy-out/energy-in) efficiency of common battery technologies, such as the lithium-ion that energizes the laptop on which I write these words, is roughly 90 percent.

The expansion of hybrid cars into the mass market and the announcement of battery-driven cars such as the Chevrolet Volt and the Tesla Roadster suggest that battery technologies may be within reach. Given the economic incentive from the oil-to-electric price differential and the depth of government support for reducing petroleum dependence, it seems plausible that, if oil prices stay high, half of all personal vehicles sold by 2020 might have partially electric power plants. Of these, most would be hybrids, a smaller number would be plug-hybrids and an even smaller fraction would be pure electric vehicles.

Cheap batteries are not the only way to electrify transportation. It is technically possible to transmit power to moving vehicles. Roads can be built with embedded wire loops that transmit power to vehicles by electromagnetic induction. As with the development of other new vehicle fuels such as hydrogen, the development of electrified roadways faces a chicken-and-egg problem: drivers will demand that their electric cars be "refuelled" at least as conveniently as they fill the gasoline-powered vehicles they drive today. Yet it makes no economic sense to build a massive roadway or refueling network for a small vehicle fleet.

But the spread of electrified vehicles might open a new pathway for the economic deployment of electric roadways. Suppose hybrid vehicles spread widely, accounting for more than half of the vehicles on the road, and that short-range electric cars gain a small but growing market share. It might then be possible to introduce electrified roads on a limited basis, and make an economic return by charging cars as they drive. Work on electrified roadways has tended to assume that vehicles would need roadway power all the time, but the rising capacity of batteries means that it may be possible to have only portions of major highways and traffic arteries electrified and yet still supply power that allows all electric vehicles to travel unlimited distances.

Moreover, we are not faced with an either-or choice when it comes to oil substitutes. Electrification and synfuels are two alternative paths beyond oil. Plug-in hybrids can burn synfuels, so we can mix these options within and between transportation sectors. And we have not even considered other paths, including biofuels and hydrogen. At one-hundred-dollars-per-barrel oil, alternatives are looking pretty interesting, electric cars are poised to enter the market and large synfuel plants are being built. At two-hundred-dollars-per-barrel oil, these alternatives would begin to look cheap, and the transition would accelerate. Given the steady reduction in oil demand per unit of GDP, industrial economies can now withstand prices in the one-hundred-to-two-hundred-dollar-per-barrel range, albeit with some real pain if prices rise fast. But given the choice, of course, they would rather not, and so will transition away from oil as prices rise, limiting demand and so, in the long run, limiting oil prices.

We have ignored the social and political complications involved in the transition away from conventional oil. But that is not the point. The fact that one could develop petroleum substitutes at a rapid pace without requiring major disruptions to the operations of modern industrial economies shows that oil scarcity in itself does not pose a catastrophic threat as claimed in some discussions of "peak oil." Of course, with bad management or bad luck the security implications of oil scarcity could be horrific: one can certainly imagine scenarios in which the world blunders into war over scarce oil. But that is not necessarily our fate.

Scenario Two: Tight Carbon, Abundant Oil

Suppose oil was abundant, but we had to sharply reduce carbon emissions to limit climate risk. Could we do it? How, and at what cost might we re-engineer our energy infrastructures?

It is a difficult question. The case for urgent action to limit carbon emissions is inherently surprising given the slow-moving and uncertain link between our emissions and climate change. The worst risks of climate change are more than half a century off, and actions to cut emissions today will have little benefit over the next decades, so it is no small matter to argue that we should begin overhauling the entire energy system at a cost that will run to trillions of dollars. I will therefore present the case for taking climate risks seriously before delving into options to manage that risk.

To illustrate the range of time scales that link carbon to climate, I will frame the story around three important dates.

The first date is the end of the Eocene, not a historical date, but rather an era in the geologic record that ended about thirty-five million years ago. The Eocene was the last period in which geoscientists are confident that carbon dioxide concentrations stood above 1,000 parts per million, about four times their pre-industrial value of about 270 parts per million. The Eocene climate was far warmer than today's. Crocodilians walked the shores of Axel Heiberg Island in the present-day Canadian Arctic.

If we continue on our present course, then within this century—within the lifetime of my children—human actions may well push carbon dioxide concentrations to levels not seen since the Eocene. Global emissions of carbon dioxide now exceed thirty-two billion tonnes per year, an average of five tonnes per capita. An average Canadian is responsible for twenty tonnes, four times the global average. If twenty tonnes sounds like a lot, it is. The mass of carbon dioxide we dispose of in the atmosphere dwarfs the mass of garbage we send to landfills, and now exceeds all human-driven material flows, including the gigantic movements of mine overburden. We throw away forty times more carbon dioxide than we do garbage. If carbon dioxide were a smelly mass that had to be pushed around with bulldozers, we would have dealt with the problem long ago.

The root of the climate threat is that humanity is moving carbon from deep geological reservoirs to the biosphere approximately a hundred times faster than the corresponding natural process in which carbon from deep in Earth's crust escapes from locations such as volcanic vents. While significant uncertainty remains about the climate's sensitivity to increasing carbon dioxide and about the attribution of recent warming to the historical increase in carbon dioxide, the extraordinary human acceleration of the carbon cycle is an undisputed fact.

We cannot accurately predict the results of the experiment we are performing upon our planet, but it is safe to say that, if we continue on our present course, we are committing our children to climate changes that will be extraordinarily rapid compared to those humanity has experienced over the ten millennia since the invention of agriculture, during which carbon dioxide concentrations stayed within about 5 percent of their pre-industrial average. It is plausible, for example, that if we were to continue emitting carbon dioxide at the current rate, the sea level would rise more than five metres within a few centuries. A sea-level rise of just a few metres is enough to dramatically alter coastlines as seen from space. Low-lying areas such as Florida and the Fraser River Delta would be innundated.

There is nothing wrong with the Eocene climate; there is no inherent reason we should prefer our crocodiles in the Florida Keys rather than on Axel Heiberg Island. The climate risks come from the rate of change, not because the current climate is some magic optimum for life. Our infrastructures, our crops, the very locations of our coastal cities have evolved for the current climate. The slow adaptation that has anchored us to the current climate puts us at risk if climate changes fast. The climate has varied for billions of years, and would keep changing without us, but on our current high-emissions path, the rate of climate change over the next century will likely be at least ten times more rapid than humanity has experienced in the past millennia.

While it is beyond the ability of social science to predict accurately the economic and social repercussions of rapid climate change, and while there will be some sectors or regions that will benefit from climate change, it is very likely that the losers will outnumber the winners. Economic analysis of climate impacts typically focuses on the aggregate economic damages which are thought to run to a few percent of GDP. Underneath this aggregate, some regions and industries have huge gains while others suffer huge losses, yet in focusing on the aggregate economists implicitly assume that the winners will compensate the losers. More likely, it is the sociopolitical tensions arising from this rapid reshuffling of the deck that will pose the largest risks.

The second date is 1965, the year in which the first scientific warnings about how the climate would be affected by fossil fuel combustion reached the ears of the world's most powerful politician. The report that President Johnson received from his scientific advisors related the same essential facts we know today: carbon dioxide concentrations are rising due to fossil fuel combustion; if left unchecked, substantial climate change is to be expected within a century; the exact amount of change and its consequences for humanity are uncertain.

I mention the 1965 report to emphasize the constancy of scientific understanding and to illustrate the absurdity of claims made by climate skeptics such as Canada's "Friends of Science." Such skeptics would have you believe that the climate story has been constructed by left-leaning Environment Canada scientists, or that the basis for concern depends on supercomputer models or on arcane details of the reconstruction of recent temperatures (the "hockey stick" debate).

Every one of these three assertions is nonsense. One of the best climate-research groups is at Lawrence Livermore, a top U.S. weapons lab, where atmospheric modelling expertise was built early to forecast where the fallout plumes would go in a hot war. Many of the Livermore scientists are not left-leaning enviros, and would be strongly motivated to report serious errors in the climate science to the U.S. government if they found them. Moreover, the scientists reporting to Johnson correctly forecasted warming, even though the globe was then cooling. So, contrary to common assumptions, the estimates of the climate risk have little to do with year-by-year temperature shifts, just as estimates of oil depletion have little to do with short-term movements in the price of oil. Finally, the basis for predicting a temperature rise in response to rising levels of carbon dioxide ultimately rests on foundations like the physics of atmospheric radiation, which has been known for more than a century, and does not depend on the details of the latest computer simulation. Using that physics in the 1890s, Svante Arrhenius was able to make a prediction for the warming due to carbon dioxide increases that fell squarely within the range of current computer simulations.

More than four decades have passed since scientists first sounded the alert about carbon-climate risks, and more than a decade has passed since climate change emerged on the international political stage with the negotiation of the Framework Convention on Climate Change and the Kyoto Protocol to that convention. Yet we have accomplished almost nothing. Far from reducing emissions, over the past decade, the annual growth of emissions has accelerated from just over 1 percent to more than 3 percent per year. If we are to reduce carbon emissions, what matters are actions, and meaningful action on this topic has been all but nonexistent. Sadly, perhaps the most measurable impact of the international negotiations has been the carbon dioxide emissions from participants' air miles.

The last date to consider is 2035. If the growth of carbon dioxide concentrations continues unabated, it will exceed 450 parts per million by about 2035. This concentration is roughly equivalent to a doubling of pre-industrial carbon dioxide concentrations when one includes the influence of other greenhouse gases such as methane and nitrous oxide. A carbon dioxide doubling has long been the benchmark for the level of greenhouse gases that would cause substantial climate change, posing real though uncertain risks to human society and the natural environment.

In the slow-moving world of carbon cycles, climate change and energy infrastructure, 2035 might as well be tomorrow.

This fact is the centrepiece of the argument for urgent action to restrain carbon emissions.

I will return to the theme of uncertainty in the conclusion of this essay, but for now let us assume that we wish to reduce carbon dioxide emissions in order to manage the climate risk. What would we do?

As with the challenge of oil scarcity, I will focus on the technology and economics of cutting carbon dioxide emissions in a hypothetical world of omniscient and benevolent central planners, and leave discussion of real-world politics for the concluding section.

Because the carbon we emit persists in the atmosphere for centuries, stabilizing the climate means that emissions must be essentially eliminated. Like water dribbling into a bathtub, even a slow flow will increase the level to overflowing. It is the carbon in the air that causes warming, not the current year's emissions, so shutting off emissions instantly will not eliminate the climate risk, just as turning off the tap will not drain an overfilled tub.

Minor cuts in our emissions intensity will simply not do the trick. While it might be possible to achieve modest reductions in emissions through reductions in demand or improvements in efficiency, deep reductions require a different strategy. The only way to retain the myriad benefits that energy use grants us—from mobility to communications to plentiful food—while at the same time eliminating emissions is to change the way we make energy. We must decouple the production of energy from the emission of carbon to the atmosphere.

Cutting carbon emissions need not, therefore, mean cutting energy use. Indeed, one can imagine futures where emissions are cut and energy use accelerates.

Decoupling energy from carbon means either switching to non-carbon energy sources such as solar, wind, biomass and nuclear power, or finding ways to use fossil energy reserves without leaving the carbon in the atmosphere by capturing carbon dioxide from energy transformations and disposing of it safely underground.

None of these energy options is a panacea. At the scale of modern energy use, all energy technologies carry environmental risks and their implementation will have profound and far-reaching consequences for human societies. Large-scale use of biomass energy such as corn or wood carries profound environmental risks because land must be diverted from other human uses or appropriated from nature. At their root, these risks arise from the mismatch between the scale and intensity of human energy use and the diffuse energy production by biological systems. To a lesser extent wind, hydro power and some other renewables will produce risks for the same reasons: our civilization now uses energy at such a rate that diverting it from small or diffuse natural flows will necessarily cause significant environmental harm. Solar power is more plentiful and more intense, so it seems plausible that it could satisfy all human energy needs without unacceptable side effects. However, the high cost and intermittency of solar currently stand in the way of large-scale use.

Nuclear power and fossil fuels with carbon dioxide capture and storage pose risks of a different kind. These sources generally produce far more energy per unit of land than do renewables, so the diffuse environmental impacts can be less serious (for a given amount of energy production) than the impacts of low-intensity renewables. The flipside of the coin is, of course, that these technologies pose some acute risk of accidents, and that the risks extend far into the future. Nuclear power poses a whole other class of risks because of its connection with the proliferation of nuclear weapons.

If well managed, all of these technologies could provide energy with substantially less overall environmental risk than the current energy system. There are no risk-free energy systems, and in considering the risks of these low-carbon-emission options, we must bear in mind the extraordinary environmental impacts—from climate change to air pollution—of our current energy system.

While there is no certain path to a carbon-free energy future, there are many steps we could take today that would take a substantial bite out of our carbon emissions. In most countries, a serious attempt to cut emissions would likely start in the electricity sector, the source of more than 40 percent of the world's carbon dioxide emissions, almost all of which are from coal-fired power plants.

There are three major options for producing near-zero-emissions electricity that could be implemented on a large scale at costs that are less than twice as expensive as current generation technologies: wind power, coal with carbon dioxide capture and storage, and nuclear power would each allow us to make rapid and deep reductions in emissions from electric power systems. Technologies like hydro, solar-thermal and geothermal can be competitive where there are rich resources, fast-flowing rivers, geothermal hot spots or sunny deserts, but they cannot now be widely deployed at costs comparable to the big three.

When transmission, distribution and marketing costs are factored in, generation costs amount to between one-third and one-half of an average electricity bill, so even if decarbonizing electricity doubled the cost of generation, it would increase a typical bill by less than 50 percent. Costs might well be lower if regulations were implemented slowly enough to allow new technologies to reduce generation costs.

No one wants their electric bills to increase, but one does not need a fancy economic model to understand that if carbon controls were implemented with deliberate haste, our society could easily afford such an increase in electricity costs without significant disruption. We can afford to protect ourselves from climate risks.

It's easier to squeeze the carbon out of electricity generation than it is to squeeze it out of transportation. This is because in electricity generation the costs of cutting emissions are less per tonne of carbon and because electricity generation technologies can be freely mixed, albeit with some constraints. The chicken-and-egg linkage of cars to their refuelling infrastructure makes it harder to mix novel fuels such as hydrogen or electricity. Nevertheless, there are credible routes to making deep cuts in the carbon emissions from transportation beyond what can be achieved by efficiency improvements alone.

For personal vehicles, electrification may well be the leading option, though its effectiveness depends on the decarbonization of electricity. It seems unlikely that hydrogen can be a serious player in the transportation system beyond minor niches. Both its relatively poor end-to-end energy conversion efficiency and the inherently poorer fuel handling characteristics make hydrogen's large-scale use unlikely.

Despite the challenges of managing the environmental footprint of biofuels, they remain a serious contender. There are interesting options beyond incremental improvement of the current fermentation-based biomass-to-liquids technology. One such example would be to use biomass as a feedstock in a synfuel process much like that used for coal, while providing external inputs of process heat and hydrogen from a non-biomass source. These non-biomass sources could include nuclear power, coal with carbon capture or a future solar-to-hydrogen technology.

The input of external energy makes these hybrid biomass systems very efficient at converting the carbon in raw biomass to carbon in the final hydrocarbon fuel. Early studies suggest that this method could roughly triple the biofuel yield per unit of biomass input, making it far more plausible that limited biomass resources could be stretched to provide a substantial reduction in transportation emissions. This option does not depend on some undiscovered technical magic. It simply means using biomass as the carbon source for making hydrocarbon synfuels using energy supplied by some more abundant energy resource, rather than using biomass as the source of both carbon and energy, as is the case with current biofuel production.

The hydrocarbon fuels (that is, primarily gasoline and diesel) produced by such a system would still contain carbon, and would have all of hydrocarbon's advantages such as high-energy density and compatibility with existing energy systems, but the carbon content of the fuel would have been drawn from the atmosphere when the biomass feedstock was grown, so the fuels would be, paradoxically, carbon-neutral hydrocarbons. The carbon would simply be a reuseable energy carrier, like a beer bottle that is recycled hundreds of times. The amount of high-energy carbon produced as hydrocarbon fuel would be matched by the input of low-energy carbon in the plants (such as agricultural and forest product wastes) that provided the biomass feedstock.

It is also possible that carbon needed to synthesize carbon-neutral hydrocarbon fuels could be captured directly from the atmosphere by a technological carbon-scrubbing process, "air capture," allowing the production of carbon-neutral hydrocarbon fuels without use of biomass feedstocks. While I am personally involved in the development of such technologies, I would be the first to concede that they are a long way from commercialization. Also, as with the biofuel synthesis described above, making a hydrocarbon fuel from carbon dioxide would require large inputs of energy and hydrogen.

Restraining emissions will not be free. Contrary to some overly optimistic assertions, the economic side benefits from developing and deploying low-emissions technologies, while real, will not eliminate the costs of action. But these costs are manageable. In a world with omniscient global governance, the climate problem would be easy enough. With wise investment, it seems likely that one could transition to a zero-carbon economy in less than a century at a cost of 2 percent of GDP—comparable to the amount we spend on the military and far less than we spend on education or health care. This does not mean reducing growth rates by 2 percent, which would have a catastrophic economic impact; it simply means forfeiting a few years of growth out of the next half century. In this light, climate change seems like a far lesser problem than the challenge of managing nuclear or biological weapons in a world in which war is still the ultimate method for settling disputes between nation states.

Finally, it's worth mentioning that Canada's greenhouse gas

emissions are exceptional in many respects. Because we have so much carbon-free hydro power, electricity accounts for only 20 percent of our emissions, about half as much as in a typical industrial economy. Since electricity generation is typically the easiest sector in which to begin cutting carbon emissions, the fact that we have a relatively low-carbon electricity sector makes it comparatively harder to cut Canadian emissions than it would be in a more typical industrial economy. This fact combined with the fact that Canada has a rapidly growing and resource-intensive economy makes it harder to cut emissions here than it would be in the U.S. or Europe. This is one of the reasons why Canada's choice of Kyoto target was particularly challenging, though this fact is not, of course, an excuse for inaction.

Oil and Carbon in an Uncertain World

Isolating the oil and carbon challenges is a convenient fiction, but in reality, of course, we face both simultaneously along with a host of other related challenges, from air pollution to the seemingly irresistible spread of nuclear weapons.

Before turning to the economic and political factors that shape our responses, it's worth considering some of the technical challenges of jointly managing carbon constraints and oil scarcity. While efficiency improvements reduce emissions and oil consumption simultaneously, the development of substitutes such as synfuels or electric vehicles does not. Just the opposite. Coal-to-liquids—one of the easiest large-scale petroleum substitutes—has life-cycle emissions that are roughly twice as large as conventional oil. Accelerated development of extra-heavy oil such as that found in Canada's oil sands will also increase emissions, though several new oil sands operations now have well-to-wheels emissions within 20 percent of conventional oil. Even electric vehicles will not provide significant carbon benefits if the electricity that charges their batteries is supplied by coal.

Moreover, oil scarcity will tend to focus money and political attention on the transportation sector, making it harder to sustain the investments needed to lower the carbon footprint of electricity, the most efficient place to focus if carbon were our primary challenge.

During the recent oil-price surge, for example, many commentators have argued that high oil prices made carbon policy unnecessary. This is wrongheaded for at least two reasons. First, if, as seems likely, oil scarcity drives a switch to extra-heavy fuels and coal, high prices will increase emissions; and second, most carbon emissions worldwide, though not those in Canada, come from coal, and it has not seen a similar rise in price. High prices and oil scarcity will not solve the carbon problem; on the contrary, the high-carbon emissions associated with a shift to ultra heavy oil and coal-to-liquids as we "scrape the bottom of the barrel" will counterbalance the demand reduction spurred by high prices.

The upshot of this interplay between oil scarcity and climate risk is that we cannot hope that politicians will be able to craft wise energy policy if we lack agreement on the goals which that policy must address. It is naïve to imagine that there is a generic "good" energy policy. Policy must be designed to address specific goals, and we must agree on the relative importance of potentially conflicting goals such as supply security and climate security before we can make sensible energy choices.

As a first step towards the real world, consider the differences between the oil and carbon challenges that arise from the fact that oil is a commodity while climate protection is a global public good.

Meeting either challenge requires a massive investment in infrastructure in order to reshape our energy system so that it can supply the energy services we need—such as mobility, illumination and communication—while eliminating the need for oil or the emission of carbon. We have technologies near at hand that could provide a substantial step to resolving either problem at costs that would make a scarcely discernible dent in our economic fortunes, yet the central distinction between the two is economic, not technical.

The transition away from oil will be aided by the market's magic. As oil becomes scarce, prices will rise and it will be in the self-interest of individuals, corporations and nations to find substitutes. While governance and collective action will be needed, they will be greatly aided by self-interest that tends to make oil scarcity self-correcting.

Unfortunately, naïve self-interest cannot be relied on to help solve the climate problem. For carbon as for oil, the costs of technical substitutes are local, yet for carbon the benefits of reduced climate risk are spread globally. It is in each nation's direct self-interest to leave their emissions unchecked while exhorting others to reign in their emissions. In the language of economists, climate stability is a global public good.

Consider our family's recent purchase of a super-efficient furnace. We bear the full cost (net of fuel savings) of our climate-friendly hardware, while the benefits of its reduced emissions are spread worldwide; and worse, because of the slow dynamics of the carbon cycle, the benefits are nearly zero over the rest of my working life. The global climate benefits from our furnace, but only as the reduction in emissions gradually produces an (infinitesimal) reduction in climate change that extends more than a century into the future. It's a lousy deal. We can only justify it by the hope that our family's purchase will spur others to make the same choice, thereby amplifying the climatic benefits beyond those achieved by our furnace alone.

Humanity has solved global public-goods problems before, the most notable example being the near elimination of ozone-destroying chemicals from the global marketplace. The trick involves national and international mechanisms to reduce the incentive for free riders who would let others shoulder the burden of reducing emissions. But there is no magic formula to ensure success. Moreover, despite the grand statements issued by the current round of climate negotiations, current attempts to cut carbon emissions are more like the "phony war" that preceded World War II than real actions of the kind and magnitude necessary to attack the problem.

Uncertainty makes both the climate and oil problems harder, and it is uncertainty in combination with the public-goods problem that makes the carbon-climate problem so dangerous.

Substantial uncertainty about the climate's response to our meddling will remain unresolved for decades. At the low end of current estimates, the global average temperature would rise one degree Celsius, a bit more than its rise since 1900, while at the high end, the average temperature would rise about five degrees, a global climate shift more than half as large as the shift from the ice age to the present day.

If the climate risk could be quickly eliminated, we could wait until the uncertainty about the degree of climate risk was resolved before taking action. Upon discovering the extent of the climate sensitivity to carbon, we could then choose an appropriate response, one that balances the cost of response against the environmental benefit.

Unfortunately, we cannot sensibly postpone action until we know the exact degree of climate risk. The problem is the enormous inertia between our decision to respond and the benefits of that response. The inertia arises from the long lags in the climate's response to our carbon emissions—more than half of the carbon we emit today will be retained in the atmosphere a century hence. Similarly, our carbon-emitting energy infrastructure is not going to be replaced overnight. The turnover of energy infrastructure is relatively slow; power plants are not replaced yearly like iPods, they often last half a century.

It is the combination of high uncertainty with high inertia that makes the climate problem particularly dangerous. We cannot wait to find out if the climate dice have rolled against us before we act. Planning our response around the most likely outcome is reckless overconfidence. However, we seem particularly unable to manage problems that combine high uncertainty and high inertia. The tendency to procrastinate is just too strong. This fact, combined with the climate's status as a global public good—the benefits of cutting emissions are spread globally while the costs are felt locally—goes a long way to explaining humanity's near-total failure to act in the face of the climate threat.

Uncertainty in the face of the climate threat should not be a justification for inaction any more than uncertainty in the battlefield is a reason to delay decisions about the movement of troops. We should make decisions in the face of uncertainty, accepting that the consequences of climate change may turn out to be more or less severe than our best guess. We must hope for the best while laying plans to navigate the worst.

So, on the one hand, we have an energy problem that, while potentially dangerous, is largely self-correcting by the normal market incentives that arise from scarcity. As oil dries up, the self-interest of nations, firms and individuals will automatically drive the implementation of substitutes from coal-to-liquids to electric vehicles. That underlying self interest does not justify complacency. Navigating the oil transition will be dangerous and clear-headed actions by governments will be needed to provoke energy innovation and to manage the energy-system transition; but for the oil transition, government action will be pushing in the direction that the economy naturally "wants" to go.

On the other hand, we have a planet-altering problem in which current actions pose growing risks to future generations. For climate change, the self-interest of nations, firms and individuals will work to drive measures to ease adaptation to the changing climate since the benefits of adaptation can be captured locally where money is spent, whereas cutting carbon emissions demands coordinated actions to secure a global public good. Far from being self-correcting, this problem may well be self-reinforcing. If actions to limit emissions continue to fail and climate change accelerates, then governments may focus their efforts on adaption, abandoning a coordinated attempt to control emissions with potentially disastrous consequences.

The oil and climate challenges are both rooted in our energy system. Abundance of fossil and other energies enables the energy substitution that is the path beyond oil scarcity, yet it is the very abundance of fossil fuels that drives the climate threat. Fossil abundance means that low-cost high-carbon fuels such as coal and ultra-heavy oil tend to out-compete the lower carbon alternatives and create the carbon emissions that drive climate change.

Climate is therefore the arena in which government action is by far the most important. The innovations and investment that limit carbon emissions will be (mostly) made by firms and individuals, but their efforts cannot be harnessed unless governments act to put a price on carbon and to build international mechanisms that trigger the myriad small investments that must be made to limit emissions worldwide. Investments that will be repaid many times over if we are able to leave our children with a climate like that in which our civilization evolved.

J. David Hughes
THE ENERGY ISSUE:
A More Urgent Problem than Climate Change?

J. David Hughes is a geoscientist whose career spans nearly four decades in the energy sector, including thirty-two years with the Geological Survey of Canada as a scientist and research manager. He developed Canada's National Coal Inventory, which is a digital knowledge base used to determine the availability and environmental constraints associated with coal resources for conventional and unconventional uses. As Team Leader for Unconventional Gas for the Canadian Gas Potential Committee, he coordinated the publication of the Committee's latest review of Canada's unconventional natural gas potential. He has published and lectured widely on energy issues across North America and internationally. His analyses of global and North American energy issues have been presented to federal, state, provincial and municipal agencies, industry groups, professional organizations and the general public. Aspects of his analyses have also been taken up by the popular press, trade journals and other media. He is on the Board of Directors of ASPO Canada and is a Fellow of the Post-Carbon Institute. He is also president of a consultancy dedicated to research in global energy and sustainability issues.

If you watch the news or read the papers, the issue of climate change caused by mankind's profligate use of fossil fuels, and its impact on the global environment, is everywhere. Although there is considerable angst at the doubling of world oil prices in the past year and the effect of this on gasoline prices and the driving habits of North Americans, the underlying cause is generally not talked about—the fundamental fact that the fossil energy we all rely on is a nonrenewable, one-time resource. Climate change is a serious issue that must be addressed, but the limitations in our ability to grow energy supplies to meet growing demand, given the correlation between energy consumption and economic growth, will probably have a far more severe impact on our lives than climate change in the near future. Fortunately, several of the solutions to climate change also directly address the energy sustainability issue.

Certainly there is a lot of talk about massive finds in deep-sea oil wells off the coast of Brazil, which could contain as much as forty-one billion barrels of oil. A recent report also indicates that the circum-Arctic region may hold as much as ninety billion barrels of oil. In a world thirsty for oil, this is all good news. But we need to put these discoveries into perspective. If we consider that the world burns thirty-one billion barrels of oil each year, the Brazil find represents sixteen months of world oil consumption and the entire circum-Arctic region perhaps three years. There is also talk about speculators unhelpfully bidding up the price of a barrel of oil, and about Saudi Arabia's promise to open the taps a little to increase supply, even though the Saudis are producing less oil today than they were three years ago. One of the reasons politicians, television news anchors and newspaper columnists are so reassuring about our energy future is that the people *they* get their information from are just as bullish.

There is no shortage of individuals, groups and agencies analyzing future energy supplies, many of which are focused on oil. Two agencies regularly produce outlooks for the world for all energy sources: the U.S. Energy Information Administration (EIA), which is the statistical arm of the U.S. Department of Energy, produces an International Energy Outlook (IEO) each year; and the International Energy Agency (IEA), responsible to the Organisation for Economic Co-operation and Development (OECD) countries, produces a World Energy Outlook (WEO) on an annual basis.

Two things about these reports make them worth mentioning here. First, they are what politicians and governments use to cobble together a credible view of the world's energy future. Second, both of these agencies, the EIA and the IEA, invariably paint a view of the future that is barely distinguishable from the past. That is because consumption growth trends from the past are projected into the future—

and it is *assumed* that energy supplies will be available to fill in the gap under the demand projections. The concept that non-renewable resources are finite at some level and that *rates of supply* are approaching their maximum for geological and geopolitical, not economic, reasons does not factor into their analysis. Energy policy and planning is built around the assumption that things will continue as they have gone for the past few decades.

These agencies believe that current high and volatile energy prices are no more than a temporary aberration, and that technology, the market and the laws of economics will soon sort things out and prices will fall back down. Inevitably, so they think, business as usual will continue, unhindered by the reality of finite resources. The reference-case forecast of the U.S. Energy Information Administration's IEO for 2008 states that world energy consumption will grow by 44 percent through 2030,[1] and the most recent WEO from the International Energy Agency states that world energy consumption will grow by 45 percent through 2030.[2] But a quick, back-of-the-envelope calculation suggests that we should question these reports' assumptions: if their calculations are correct, world energy consumption in 2030 will be five times the amount of energy consumed by the world in 1960— around the time crude oil discovery peaked. Is there really that much energy out there to use?

We might be more inclined to go along with these assumptions if both the U.S. Energy Information Administration's and the International Energy Agency's forecasts of energy prices and supply didn't have such dismal track records. For example, in the just-released IEO for 2008, the EIA reference case is for oil prices to fall to $56 per barrel by 2016 and to be $70 per barrel in 2030 (in 2006 U.S. dollars).[3] Although oil prices (along with the prices of many other commodities) collapsed in the fall of 2008 from a high of nearly $150 per barrel due to the unwinding of the global economy and the resulting uncertainty about future demand, world oil consumption is still forecast to grow in 2009, and this respite in prices is likely to prove temporary. The IEA's WEO for 2007 is no better, forecasting $62-per-barrel oil in 2030 and natural gas prices of less than $8 per million British thermal units in 2030, even in Japan, where spot prices of $20 per million British thermal units are being paid today (though the latest WEO, released in November 2008, forecasts $100 a barrel oil in 2015 and $120 in 2030).[4] The IEA's forecast that coal prices will be flat 2015 has already been proven wrong; seaborne thermal coal prices for power generation have doubled and metallurgical coal prices have tripled in the past 18 months. It would appear that reliance on the dismal science (economics), rather than the more down-to-earth geological and engineering realities, distorts our sense of what is possible in terms of exploiting finite resources.

Meanwhile, those analysts who acknowledge the finite limits of hydrocarbons are projecting oil prices to hit $225 per barrel by 2012 and over $300 per barrel by 2020. The EIA and IEA forecast a smooth ramp up of global oil production from 86 million barrels per day at present to between 106.4 and 118.2 million barrels per day by 2030 (IEO 2007,[5] WEO 2007,[6] WEO 2008,[7] IEO 2008[8]) when many serious and highly credible analysts insist it will be difficult to make it past ninety million barrels per day because of the difficulty, cost and geological limits of growing production. Barring miraculous new discoveries, which are unlikely given the intensive global exploration effort to date, I think the production of much more than ninety million barrels per day will be extremely difficult to achieve.

The economic meltdown and credit crisis of the fall of 2008 have resulted in the cancellation or deferral of many investments in new production infrastructure, which will be felt in a few years' time, when these projects would have come online. The IEA's latest WEO, while projecting growth in world oil supply to more than 106 million barrels per day in 2030, points to an average 6.7 percent yearly decline rate of the world's oil fields that are past peak production.[9] This would mean that the equivalent of six Saudi Arabias' worth of new supply will have to be brought online by 2030 to make up for depletion and to meet projected demand growth—an outcome akin to a lottery win in my opinion.

It seems the news is not all good after all.

Some Background

As a geologist working for the Geological Survey of Canada and in the private sector on energy for more than three decades, I became concerned about what I call the "Energy Sustainability Dilemma" many years ago. This dilemma arises from the current economic paradigm based on continuous growth, which is closely linked with growth in energy consumption. The "dilemma" is posed by the end of our ability to grow or even maintain current rates of energy consumption due to geological and geopolitical fundamentals. Do we press on with business as usual at all costs until we hit the wall on energy, or do we recognize the unsustainability of this paradigm and adopt a new paradigm of living within the means of our energy options? I am certainly not alone in my concern that we simply will not be able to go on using energy in the way we have become accustomed to (just type "peak oil" into a search engine), but the lack of widespread understanding of this incredibly important issue, not just amongst the general public but also amongst many academics, most economists and, outwardly at least, most politicians, is astounding to me. Our integrated energy system is exquisitely vulnerable to disruption, and it's built around assumptions that are proving to be illusions. All of our eggs are stacked in one very precarious basket.

To get a sense of just how precious fossil fuels are to our way of life, consider the following two facts. Hydrocarbons (by which I mean oil, natural gas and coal) made up 89 percent of our primary energy consumption in 2007, the balance being a cocktail of nuclear energy, large hydroelectric energy and other renewables. Now consider that a barrel of oil is equivalent to more than eight years of human energy.[10] At minimum wage, the human energy equivalent of a barrel of oil would be $138,363 and the human energy equivalent of a gallon of gasoline would be $3,294. That is to say, more than 99 percent of our energy comes to us nearly free, even at current prices.

Hydrocarbons are truly the "elixirs" of modern life, and yet we hardly pause to think about them, preferring to assume that there will always be more just beneath our feet. But even the most commonsense appraisal of this way of looking at our civilization's fuel shows that hydrocarbons cannot possibly be infinite. They represent the legacy of five hundred million years of solar energy, preserved in plant and animal remains. They are "fossilized sunshine," solar power distilled to an energy density by weight and volume that makes them incredibly useful and precious. And they are made all the more precious by the fact that, on a human time scale at least, they are non-renewable.

This touches on what I mean when I talk about the Energy Sustainability Dilemma (something I am often called upon to do). We need energy to run our global civilization, that much is obvious. Global per capita energy consumption has risen eightfold since 1850 (roughly the history of industrial civilization). And this increase has come about not by one type of fuel replacing its predecessor, but by one type of fuel adding to its predecessor.

Prior to the advent of hydrocarbons, energy was provided primarily by renewable biomass (mostly wood), human labour and draft animals. The energy system was essentially sustainable, as it was based primarily on renewable solar energy, rather than on Earth's one-time legacy of fossil fuels as it is today. The sun nourished the trees and the crops, which, when harvested, nourished the labourers and draft animals—the energy system thus was in balance with the daily input of solar energy, and hence was sustainable.

By 1850, when Earth's population was just over 1.2 billion, 82 percent of primary energy was still provided by biomass; the balance came from coal. Coal consumption has risen ever since, and by 1910 amounted to 70 percent of primary energy consumption. Though coal's share of our total energy consumption has fallen since then, the average global citizen consumes the same amount of coal per capita today as in 1910, and the total amount of coal consumed has increased by 4.2 times since then.

Although oil was first tapped in 1858, its use didn't really take off until after World War II. Growth in the consumption of energy from oil was not at the expense of coal—it served only to increase the total energy consumption of the average global citizen. In 1950, 50 percent of primary energy was still provided by coal, 25 percent by oil. Today, oil consumption accounts for 34 percent of primary energy—and we are consuming 7.7 times more of it than in 1950.

The growth in consumption of natural gas followed a similar pattern to that of oil. At first, energy seemed so limitless and natural gas so difficult to transport that it was considered no more than a waste product associated with the production of oil. It was flared (that is, burnt off into the atmosphere). But natural gas gradually became an increasingly important contributor to primary energy consumption and, as with oil, served to increase the total energy consumption of the average global citizen, instead of replacing the consumption of coal or oil. In 1960, natural gas provided 13 percent of primary energy consumption. Today we are consuming 6.1 times as much natural gas as in 1960, and it now provides 23 percent of primary consumption.

The eightfold growth in global per capita consumption of energy we have enjoyed since 1850 is because of the exploitation of non-renewable fuels—oil, natural gas, coal and uranium. Although renewable energy, such as biomass, hydroelectricity, wind, solar and geothermal, has increased in real terms since 1850, the average global citizen consumes the same amount of renewable energy per capita today as in the middle of the nineteenth century. We can't go *back* to using renewables—we're already using them as much as we ever did. But now, 89 percent of the energy we use to heat our homes and to fuel our cars, airplanes, factories and power stations is provided by non-renewable fuels. Each of these non-renewable fuel sources will have a peak in available supply, probably in the first half of this century. That's why understanding these future supply issues is crucial to planning for civilization as we know it. I will deal with each hydrocarbon fuel source separately, but first, there is one very important concept to be clarified.

The Concept of Net Energy

Not all hydrocarbon resources are created equal—so when we hear that there are so many billion barrels of oil in the tar sands or in this or that deep-sea find, we're getting only half of the story. Maybe even less.

It costs energy to get energy, and the whole point is to get back more than you put in; this is called Energy Return on Investment (EROI). The difference between the energy expended to get hydrocarbons and the energy contained in the hydrocarbons is termed "net energy."

The Ghawar Field in Saudi Arabia, the largest oil field in the world, was discovered in 1948 and brought online in 1951. It is responsible for 50 percent of Saudi Arabia's oil production and has had a fabulous payback in terms of EROI— vertical wells drilled in the 1950s yielded twenty thousand barrels per day or more, for a net energy payback of greater than 100:1. That is, for every barrel equivalent of energy invested, one hundred barrels came gushing back. That is a return any investor would jump at.

Things aren't quite the same today, however. New discoveries, such as the Jack-2 discovery in the Gulf of Mexico and the Tupi find off the coast of Brazil, are in much more extreme settings, requiring rigs capable of drilling to more than twenty thousand feet in five thousand or more feet of water. Oil from these discoveries is very difficult to produce and results in a much lower net energy profit. Even the biggest of the most newsworthy new discoveries are quite modest in size and in EROI compared to the big oil fields of yesteryear. And even supergiants like Ghawar aren't what they used to be. Oil no longer gushes to the surface; instead, engineers deploy the latest in oil-field technology and inject seven million barrels per day of water into Ghawar to pressurize the reservoir, all of which diminishes the EROI of each barrel they coax out (about five million barrels per day).

The unconventional oil and oil substitutes that were heralded as the future of our hydrocarbon-based civilization have an even more sobering tale to tell. Tar sands and shale oil are at the meagre end of the diminished EROI spectrum—mineable tar sands are about 6:1, in situ tar sands development through Steam Assisted Gravity Drainage (SAGD) is about 3:1, and shale oil has not yet been extracted at a net energy profit on a large scale. The corn-ethanol boom in the U.S. has an EROI somewhere between 1.3:1 and 0.79:1, meaning that, depending on whose studies you believe, it is somewhere between an extremely poor source of liquid energy and an energy sink. Although sugar cane ethanol in

Brazil and palm-oil biodiesel in Indonesia have a higher EROI than U.S. corn, there are other ecological consequences of large-scale exploitation of these sources. The holy grail of biofuels, large-scale production of "cellulosic" ethanol, has been just around the corner for years and is likely to continue to be.

And in general, the worse a fuel's EROI, the larger its carbon footprint. All hydrocarbon reserves fall somewhere on the quality spectrum between the gigantic reservoirs that ushered in the oil age and the smallest and most exhausted of the oil fields more common today: they range from very large accumulations to highly dispersed, very small accumulations; from supergiant, highly pressurized light oil fields to very small, heavy oil fields and tar sands; from highly prolific gas fields to very small pools; from thick, near-surface coal seams to thin, deep, non-extractable seams. As we move down the energy chain from the highest-quality hydrocarbon resources to the lowest, the amount of energy invested in the exploration and development effort increases. We are expending more and more hydrocarbon energy in the act of obtaining more hydrocarbons instead of doing other useful work with them.

I have a diagram that I sometimes use in my presentations (see figure 1). It's a pyramid representing the hydrocarbon resource endowment of the planet, with the highest-quality deposits at the top and the lowest-quality deposits at the base. I begin by saying that hydrocarbons are essentially infinite, but then show that as we move lower into the pyramid the hydrocarbons are more dispersed into smaller and smaller accumulations that require more and

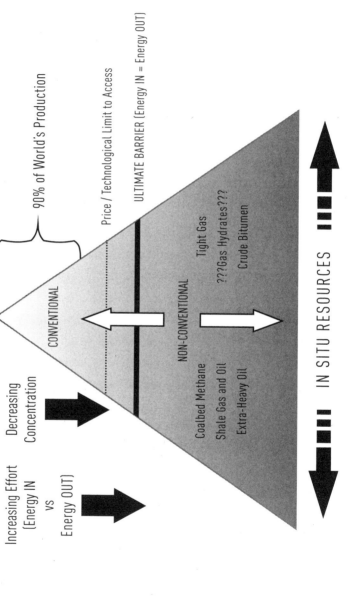

Figure 1. The concept of net energy applied to the ultimate recovery of the world's remaining hydrocarbon resources.

more energy to extract. There are two horizontal lines on the pyramid. One represents the influence of technology and price—as price moves higher and technology gets better, the line can move deeper into the pyramid, making more resources accessible. The other line is the point at which energy input equals energy output, or the line where net energy equals zero. All hydrocarbons in the resource pyramid below this line are an energy sink, not an energy source. As my colleague, who is a senior exploration manager in the Canadian oil and gas sector, says: "We are not price limited anymore in looking for prospects, it's just that there are very few prospects that justify looking at." The sobering news is that much of the hydrocarbons now in the ground are below this second line. This is oil and gas and coal that will never get burned.

The concept of net energy must also be applied to renewable sources of energy, such as windmills and photovoltaics. A two-megawatt windmill contains 260 tonnes of steel requiring 170 tonnes of coking coal and 300 tonnes of iron ore, all mined, transported and produced by hydrocarbons. The question is: how long must a windmill generate energy before it creates more energy than it took to build it? At a good wind site, the energy payback day could be in three years or less; in a poor location, energy payback may be never. That is, a windmill could spin until it falls apart and never generate as much energy as was invested in building it.

Clearly, the concept of net energy is crucial if we want to find a policy that will see us through the Energy Sustainability Dilemma. Yet the concept is lost on many economists, analysts and governments. When they are called upon to look for someone to blame for the fact that the cost of energy is nowhere near their projections, they mention terrorist premiums, ruthless speculators and conspiracy theories about rich oil companies. The net energy problem is largely forgotten. You can't break the laws of thermodynamics—no matter what the cost of a barrel of oil. If pumping that oil requires more energy than the oil itself yields, we are better off leaving it in the ground.

This cruel fact is seldom allowed to rain on the parade of the pundits who promise a rosy future for our current energy economy. They divide current consumption into purported resources, come up with resource lifetimes of a few decades or longer, and conclude "no problem." They point out that resources are abundant—after all, there are billions of barrels of oil in the ground. And they are bullish about improved drilling technology—which has, in fact, improved the rate at which oil is extracted from wells, often at the expense of the longer-term longevity of production. Moreover, they argue, alternatives will come to the rescue. But the fact is, they are not really answering the question, which is how to maintain and increase *the rate of supply*.

Non-conventional oil, such as the tar sands of Alberta, has vast in situ resources, purported to be 1.6 trillion barrels, of which 174 billion barrels is said to be recoverable. However, the tar sands are very different from conventional oil. Growing *the rate of supply* from the tar sands is an extremely capital- and energy-intensive endeavour which may, under optimistic projections, quadruple present production to five million barrels per day by 2030, or 6 percent of present world production. Non-conventional oil sources, such as the Venezuela Orinoco extra-heavy oil belt, have the same constraints—it is very time-, energy- and capital-intensive to grow supply at such sites. Yes, these sources represent an important contribution to future oil supplies, but their physical characteristics limit their ability to offset declines in conventional production. Other non-conventional sources, such as coal-to-liquids and gas-to liquids, also have *deliverability* problems, and cannot possibly offset declining conventional oil production. Biofuels, although an important incremental contribution, have limitations that have been dealt with earlier.

We're not running out of hydrocarbons yet. What we are running out of is cheap, easily extractable hydrocarbons. This is *not* a resource problem, it's a *rate of supply* or *deliverability* problem. When deliverability can no longer rise to meet growing demand owing to geological, geopolitical and declining net energy issues, we have the beginning of a civilization-defining moment. We will likely never completely run out of hydrocarbons. As my colleague Charlie Hall, who is a professor at the State University of New York, says, "We will always have enough oil for our bicycle chains."

Oil

Oil is now the number one source of energy on the planet at 34 percent of primary energy consumption, and it is not hard to see why it has become so indispensable. Oil has unique properties that make it a very convenient form of dense energy. It is easy to transport and it is easy to store and refine into fuel and petrochemical-based products. It is cheap, convenient and densely packed with energy. No wonder we have become so reliant on it—apart from the emissions that heat up the atmosphere when we burn it, oil has been a nearly unimaginable boon for humanity.

But the future is not nearly as rosy as the past. As of year-end 2007, the world had consumed 1.1 trillion barrels of oil, 90 percent of which has been consumed since 1959 and 50 percent since 1986. In other words, the more oil we burn, the faster we burn it. So the salient question is: just how much is there to pump? Ultimate recoverable estimates of conventional global oil, that is the maximum amount that can ever be recovered, range from about 2 trillion barrels (prior to 2000), to 2.25 trillion barrels,[11] to more than 3 trillion barrels in a highly controversial assessment by the U.S. Geological Survey (USGS) published in 2000. (The USGS is the geological science arm of the U.S. government.) So we have probably consumed somewhere between a half and a third of all the conventional oil humankind will ever consume. The large unconventional resources attributed to tar sands, extra-heavy oil and oil shale are often cited as reasons that business as usual can be perpetuated and U.S. energy independence secured; however, this is very likely a delusionary hope owing to the low EROI of these resources and the time and capital costs required to build the infrastructure needed to ramp up their rates of production, not to mention high carbon dioxide emissions and high-water-use problems. As a result, unconventional oil will likely only slow the rate of decline once the point of peak global *rate of supply* is reached. One thing is certain, the first half of global oil production was the easy half—the last half will be much more difficult in both financial- and energy-input terms.

The realization that we are close to halfway through our endowment of "black gold" is often called the theory of "peak oil." (Of course, it is probably misleading to name something a "theory" when it is a scientific certainty, but that is what it is called.) Peak oil is a concept that is much derided in some circles but which has attained new currency with oil prices topping $100–$140 per barrel. The notion of a "peak" was first put forward by the Shell geologist M. King Hubbert, who noticed that oil discovery within the United States (then the world's leading oil producer) had begun to drop off from an all-time high in 1930. Since it takes a relatively fixed amount of time to pump out an oil field, it seemed to make sense that oil production would follow the same downward trajectory as oil discovery after a lag time of forty years. Hubbert's speculation got a very cold reception from his peers in the world of geology, but it proved to be nearly pinpoint accurate. American oil production peaked in 1970 and has been dropping ever since. Quite naturally, people have applied his methods in an attempt to predict a global peak of production. Global oil discoveries peaked in 1965, so, given a forty-year time lag, peak oil production should be happening about now. Most guesses put the peak in the first quarter of this century; some claim it is already upon us.

It is important to note that peak oil is *not* the moment when the world runs out of oil. Rather, it is the name for the point when oil production cannot grow to meet rising demand. At peak levels of production, there will probably be half or more of remaining recoverable reserves still in the ground. However, the remaining oil will require much greater investment and more energy to recover, as this oil is located in difficult terrain (deep oceans, small pools) and in poorer-quality deposits (tar sands, oil shale and extra-heavy oil), which will severely compromise *the rate* at which we are able to extract it. As we saw with the Brazilian and Arctic discoveries, this oil simply can't be pumped (or developed) fast enough to significantly increase the global *rate of supply* because it is so difficult to get out of the ground.

This is not a theoretical problem. The world has been on an oil-production plateau since 2005. Oil producers have been struggling against the twin demons of depletion from existing fields—thought to be more than 5 percent each year—and global growth in demand of more than 1 percent per year. Two-thirds of world oil production in 2007 came from countries that are past peak production. Six member countries of OPEC (Organization of Petroleum Exporting Countries) peaked in the 1970s. The North Sea has peaked, beginning with the United Kingdom in 1999, followed by Norway in 2001 and Denmark in 2004. Mexico peaked in 2004, and the U.S., as we have seen, peaked in 1970 and is now producing at 40 percent below its best days. Certainly, production in the world outside of OPEC and the former Soviet Union is past peak, even with growth in a few countries such as Canada. A 2007 review by Robert Hirsh of current published estimates of the timing of global peak oil production results in an average estimate of a peak in 2014, with a likely more realistic estimate of 2012 when two of the more optimistic scenarios are removed.[12] (The Hirsh study was funded by the U.S. Department of Energy). Exactly when global oil production will peak is a function of future demand growth, depletion rates and the success of new developments. Should recognition of the energy problem by governments and industry result in programs for reduction in demand through conservation, efficiency, alternative sources and a rethink of infrastructure to reduce consumption, oil production could perhaps be maintained at today's levels or slightly higher for a decade or more. A global economic meltdown such as that which unfolded in late 2008 also results in a reduction in demand but is fundamentally different as it is also accompanied by a reduction in investment in exploration and production infrastructure due to lack of available capital and plunging oil prices. This situation could advance the date of peak oil considerably, as high levels of investment are crucial to offset the relentless declines from the world's oil fields. By the time investment returns to required levels as the economy recovers after a hiatus of one or more years depletion is likely to make it very difficult to meet or exceed earlier levels of production. Given business-as-usual scenarios in which demand growth is unrestrained, we will likely hit peak deliverability, albeit at a higher level, within five years or less. Let's be clear here—energy consumption cannot exceed energy supply. In other words, this is a problem that will be imposed on us whether we like it or not. We have some choice at this juncture as to what the fix will look like: it could be an intelligently managed transition or a very rough ride.

The dependence on Middle Eastern oil is yet another Achilles heel for oil-importing nations. Seventy-five percent of remaining reserves are controlled by OPEC, with most of this held by five Middle Eastern countries: Saudi Arabia, Iraq, Iran, Kuwait and the United Arab Emirates. There is less than two million barrels per day of surplus production capacity (located mainly in Saudi Arabia) in a world which uses eighty-six million barrels per day. That is a margin of scarcely more than 2 percent of daily production, and this surplus is mostly heavy sour crude, which is difficult to refine. It is not hard to see why oil prices have seen such volatility over the past few months. Geopolitical disruptions in production can upset the apple cart very easily, as can natural disruptions, such as hurricanes in the Gulf of Mexico (where oil is both produced and, more crucially for U.S. markets, refined). Moreover, rising domestic consumption within oil-exporting countries is reducing the amount of oil available for export to world markets. This is exemplified by Indonesia, one of the OPEC members that is now a net oil importer.

Earlier I mentioned a "civilization-defining moment" on the horizon, and there is no better example of a way of life on the cusp of change than the largest oil consumer on the planet: the United States, a country where the American (read "gasoline-dependent") way of life has been declared "non-negotiable." As the consumer of about a quarter of global daily oil production, the U.S. is the nation most vulnerable to supply disruptions. The U.S. imports two-thirds of its oil needs, nearly twice as much as China uses *in total.* Despite its well-publicized gestures and best intentions, the United States will never achieve its goal of energy independence as long as it relies on oil. No combination of domestic, Arctic and offshore exploration; biofuels; and oil shale development will keep America self-sufficient for long at present consumption rates.

The geopolitical implications of peak oil are of grave concern in a world addicted to globalization, mobility and the abundance of goods and services made possible by cheap and abundant energy. Oil is a globally traded and priced commodity, so even countries like Canada, which is capable of increasing the rate of production from the tar sands and meeting domestic needs, will not be exempt from the fallout of peak oil, especially in a world obsessed with growth. The world's superpower is also the world's biggest and most vulnerable oil addict. Russia, a true energy superpower, is now becoming much more assertive in the use of its energy wealth on the global political stage. China imports half of its oil and its requirements will only continue to grow with the expansion of its economy, competing with all of the other oil importers in the world for supplies. Meanwhile domestic oil consumption in oil-exporting countries continues to grow, which will probably mean a decline in the availability of oil for purchase on the world market, even if global production continues to grow. (Mexico, for example, once the number two exporter to the U.S., may cease to export oil altogether as soon as 2012.) The strains on the global oil system are now becoming evident, with rapid price increases and volatility, and conflicts in Nigeria, Iraq, Georgia and elsewhere, plus the ever-present possibility of natural disruptions, such as hurricanes in the Gulf of Mexico. Maintaining the business-as-usual paradigm of expansion in oil consumption will become increasingly difficult and eventually impossible.

Natural Gas

The fortunes of natural gas have improved a great deal since the days when it was considered no more than a waste material.

It is now a premium fuel, suited for many uses in the industrial, commercial, residential, transportation and electricity-generation sectors. It is now the number three source of energy on the planet at 23 percent of primary energy consumption. Global consumption of natural gas has nearly tripled since 1970 and is rising at 2.3 percent a year. Ninety percent of all of the natural gas ever consumed has been used since 1964, and half of it since 1990. In other words, just as in the case of oil, we are using more natural gas all the time.

Global gas consumption was about 103 trillion cubic feet in 2007, and reported global reserves were 6,260 trillion cubic feet, for a reserve lifetime of sixty years at current consumption rates—though of course consumption rates just keep rising. There is certainly a lot of recoverable gas remaining to be discovered; and total resources, including present reserves, could grow to 10,000 trillion cubic feet over time. If growth in gas consumption were to continue at the current rate of 2.3 percent per year, by 2050 remaining gas reserves from this total would only last an additional seven years. Of course, global consumption cannot continue to grow at today's rates and will level off at a *peak rate of supply*, which is likely to be some 40 percent higher than today, and to occur in the 2045 time frame.

But the real story with natural gas is that everything depends on where it happens to be found. Although North America consumed 27 percent of world natural gas production in 2007, it has only 4 percent of world reserves. Equally alarmingly, North American production peaked in the early part of this decade. That is to say, the continent faces the prospect of importing greater amounts of an increasingly

expensive fuel. The nature of the declines in natural gas production from existing wells means that ever more wells must
be drilled to keep production flat, or to stem declines. The
average decline rate of Canadian and U.S. natural gas fields is
21 and 32 percent per year respectively, without new drilling.
That means we have to drill thousands of new wells each year
just to stand still, never mind increasing production to meet
rising demand. The number of successful gas wells has tripled
in both the U.S. and Canada since 1996, yet U.S. production
peaked in 1973 and Canadian production is now falling at
about 4 percent per year. As in the case of oil, those who
think that the "market" will solve our problems by spurring
new exploration are heading down the wrong road: we are
already exploring and drilling as fast as we can.

This is what may be termed "getting closer to the bottom
of the barrel." We work harder and harder, through exploration and development, forever reducing energy returns—the
result of EROI inexorably decreasing over time—and forever
reducing financial returns. Construction of the Mackenzie
Valley and Alaskan gas pipelines will likely only provide a
small rise on the downward production profile of North
American gas. So, as with oil, the key issue is not large purported resources in the ground, it's *the rate of supply*, or deliverability, which is proving ever more difficult to sustain over
time. And all other gas fields in the world will experience this
same phenomenon as they mature.

Very recently, shale gas, an unconventional resource, has
led some to believe that it will be able to offset declines in
conventional North American gas production and increase
the overall rate of supply. This bullishness is based mainly on

rapidly increasing production from the Barnett Shale in Texas, which has grown from less than 2 percent to 7 percent of U.S. production in the past three years. The Barnett has resulted in a literal feeding frenzy by the gas industry, with more than seven thousand wells drilled, many in people's backyards in and around the city of Dallas–Fort Worth. The hype resulting from this play has been extrapolated by a few operators to other shale basins in the U.S. and Canada, causing a rise in stock prices and land values, which are now $25,000 per acre in the Haynesville Shale of Louisiana, where development is just beginning.

In reality, all shales are not created equal, and even in the Barnett, high productivities are concentrated in a relatively small area, where land prices are twenty times those of adjacent regions. Shales are highly impermeable, even though they contain gas. To produce they must have natural fracture systems and unique combinations of other geological characteristics, which are relatively localized. Production from natural fractures is stimulated by massive hydraulic fractures in horizontal wells. Although the initial productivity of these wells can be very high, the first-year decline rates are in the order of 65–80 percent, accelerating the drilling treadmill. Moreover, close well spacings are required to tap the gas, and wells are very expensive, with price tags of seven to ten million dollars per well. More sober shale gas operators in the U.S. are disturbed by the hype, and point out that offsetting the overall decline in U.S. gas production each year requires four additional Barnetts, and that nobody has the "foggiest" idea of how much gas is ultimately recoverable from shales and at what rates. In Canada, despite some shale gas development in northeastern British Columbia, overall production continues to fall.

The hope that a globalized natural gas market through Liquefied Natural Gas (LNG) will solve North America's and other countries' looming gas-supply woes and cap the price of natural gas is proving to be an illusion.[13] LNG is just too expensive, both in dollars and in energy. A very capital-intensive infrastructure is required to liquefy gas to minus-165 degrees Celsius, to transport it in special tankers and to regasify it at the destination. The energy required to do this equals 15–25 percent of the energy in the gas, depending on the transport distance. In other words, you would have to produce about 20 percent more gas in order to equal the energy yield you would have enjoyed had the natural gas not been liquefied.

It is not hard to see that the tension between supply and demand should be sending natural gas prices relentlessly higher. North American gas prices have proven to be highly volatile recently, but they are still more than double the level of a decade ago. Nonetheless, even at such highs, North American gas prices are lower than prices in Europe and Southeast Asia. This is why LNG imports to the U.S., despite expanded import capacity, have fallen as LNG cargos are sent to the highest bidder. Infrastructure bottlenecks will also play a role in limiting supply: global liquefaction capacity is only 56 percent of global regasification capacity at present and this is expected to fall to less than 32 percent by 2013.[14] (Essentially, we can regasify much more than we can liquefy.) As a result, prices on the global market will probably only rise because of competition for supply. Lack of LNG supply has already resulted in the suspension of two Canadian LNG terminals, and

the not-in-my-backyard response to the siting of new terminals in the U.S. has resulted in the cancellation of many others. The NIMBY response to siting LNG terminals is due to the
perceived danger of having ships with three billion cubic feet
of gas passing and unloading nearby. The City of Boston enu-
merates the security precautions that involve consulting with
many agencies, shutting down bridge traffic and stopping
takeoffs from Logan airport when LNG tankers arrive at the
Everett terminal several times per week. In reality, LNG has had
an enviable safety record over the past several decades, but
that does not mean it would not represent an attractive terrorist target, as studies have shown. LNG tankers have been
called "floating bombs." Thus, rather than capping the price
of natural gas in North America at low levels, as was once
thought, LNG will likely put a floor on the price of natural gas
at high levels in the future as North America becomes increasingly dependent on imported supplies.

Notwithstanding these supply issues, natural gas is a very
clean fuel compared to other non-renewable fossil fuels such
as coal, and this has made gas very attractive to planners looking to cut both greenhouse gas emissions and smog- and acid
rain–producing pollutants. There has been a large increase in
gas-fired electricity generation capacity in North America in
the past decade,[15] and this is forecast to continue, though at a
lesser rate going forward in both Canada[16] and the U.S.[17] This
may prove to have been a short-sighted plan.

For one thing, while natural gas is "clean" if you measure
emissions at the "burner-tip," it looks a lot less like a solution
to climate change when we consider the full-cycle supply-
chain emissions, particularly for LNG, which, as we have seen,

cannibalizes its own energy value. This means that you end up with more emissions per unit of useful energy. Add to this the fact that the methane in natural gas is much more potent as a greenhouse gas than carbon dioxide, and that small leaks at any stage in the exploitation, transportation and utilization process quickly cancel the environmental advantage of burning this cleaner fuel. Suddenly natural gas begins to look a lot less like a solution to our climate problems.

In any case, we can expect natural gas to become more scarce and more expensive—and knowing this can be a real advantage in planning for the future. Although certain renewable energy projects may not make sense at today's gas prices, they should be considered in the context of the likely near- and long-term price and supply considerations of natural gas. Though it is not a celebrity fuel like gasoline, which gets all the headlines, natural gas is something that we depend on. This will become more apparent when heating and lighting our homes becomes increasingly expensive, not to mention petro-chemical products and fertilizers produced from natural gas.

Coal

If natural gas is not a glamorous fuel, neither is coal. Though coal seems like an embarrassing holdover from the Industrial Revolution, it does a lot of work we would have difficulty doing without. Coal is the number two source of energy on the planet at 27 percent of primary consumption, second only to oil. Although coal is often considered a nineteenth-century fuel, 90 percent of the coal consumed by the human race has been consumed since 1910 and fully half of it since 1972. And we are burning coal faster than ever before; global consumption of coal has increased by nearly 33 percent since the beginning of 2003, and the U.S. Energy Information Administration forecasts that coal will be the fastest-growing fuel source globally through 2030. Coal is the largest fuel source for electricity generation worldwide, and metallurgical-grade coking coal is an essential ingredient in the steel-making industry. Coal is emphatically not yesterday's fuel.

Let's look at it a different way. Ours is a civilization that runs on hydrocarbons, and coal represents nearly 60 percent of what's left of the world's recoverable hydrocarbon reserves. The geopolitical risks associated with coal are less than for oil or gas as it is widely distributed, particularly in North America and Asia. It is also relatively easy to move by rail and ship, and thus higher-grade thermal and metallurgical coals are traded and priced on a global market. A great deal of coal is also consumed by power plants very close to the location of mining, in which case its price is related to the cost of extraction and not to continental or global markets. On an energy cost basis, coal is much cheaper than either oil or gas, and that is why it has a large and increasing role in the global energy mix. Although the price of thermal and metallurgical coal has more than tripled on world markets in the past few years, it remains much cheaper than oil or natural gas.

It is easy to see why we burn so much coal. It is very useful stuff. The role of coal in the global energy mix is much criticized by climate-change activists, some of whom call for its complete elimination. This is unrealistic, unless these same activists plan on spending a lot of quiet time in the dark in the near future.

Notwithstanding the abundance of coal relative to oil and gas, it is still, after all, a non-renewable resource. Several recent studies offer compelling evidence that the world's remaining recoverable coal resources, which have generally received less rigorous analysis than those of oil and gas, are overstated by as much as 100 percent. In 2007, the Energy Watch Group suggested that global coal resources could peak in supply by 2025, and that coal production in China, the world's largest producer and consumer of coal, could peak sooner.[18] Another study, by the U.S. National Academy of Sciences, also in 2007, stated that in the U.S. "recent programs to assess coal recoverability in limited areas using updated methods indicate that only a small fraction of previously estimated reserves are actually recoverable."[19] A third study, by Uppsala University, projects a peak in global coal deliverability in 2030, with a peak in China even sooner.[20] A 2007 study by Rutledge out of Caltech analyzed the production history for all major coal-producing regions in the world to determine the remaining resources of recoverable coal, and concluded that coal-resource quantities reported by the World Energy Council and others are overstated by as much as 100 percent.[21]

These studies parallel on the global scale my own analysis of Canada's coal resources. I have been responsible for Canada's National Coal Inventory for many years and have communicated with many other experts in other countries on the topic of coal resources. In situ coal resources are very large, but when the constraints of recoverability are imposed (seam thickness, seam depth, environmental contaminants, geological complexity, surface land use, etc.) these resources are a great deal smaller. Some of the most attractive deposits I worked on coming out of university in the early 1970s are now mined out. The concept that there is enough coal to replace oil through coal-to-liquids technology is a complete non-starter in my view. I think we'd best get on with the paradigm shift required to move to a more sustainable energy future. Although we are accustomed to worrying that we may end up burning too much coal, the fact is that there is probably not enough to meet forecast demand over the next two decades, and certainly not enough to entertain the notion that it can substitute for the energy we get from oil and gas.

Vulnerability and Risk Factors

One of the challenges of the coming decades will be to accurately identify our most urgent threats. Climate change is in the mind of the public and in the rhetoric of the politicians. The Energy Sustainability Dilemma is much less understood although it will probably have a more immediate and severe impact on our current lifestyle than climate change, which we will likely have to live with for centuries, due to feedback loops that have already been activated.

The fact is that we shouldn't be worried about cutting our carbon emissions—that is going to happen whether we like it or not. Studies have shown that IPCC (Intergovernmental Panel on Climate Change) estimates of the amount of remaining recoverable hydrocarbons (oil, gas and coal), which are used in its scenarios for calculating future carbon concentrations in the atmosphere, are overstated by six to seven times. In many ways, simply assuming that there are sufficient recoverable hydrocarbons to get us into catastrophic trouble with climate change implies that business as usual is possible over the long term. But that is clearly not the case, and this assumption diverts attention from the underlying problem of resource depletion, which must be managed starting now. It behooves us to intelligently manage this looming energy deficit in order to minimize what could be catastrophic consequences.

It should be apparent by now that the future is not going to look like the present. It simply cannot. For one thing, correcting the great inequity in the per capita consumption of energy in the world will push our resources far past their limit. One billion people in the industrialized world consume nearly 60 percent more energy than five billion people in the developing world. Canadians and Americans consume five times more energy on a per capita basis than the average citizen of the world, nine times that of a person in China, thirty times that of a person in India and one hundred times that of a person in Bangladesh. The developing world, however, aspires to First World levels of energy consumption. They are going to discover that this is not possible given the finite nature of both fossil fuels and other non-renewable resources.

Today China consumes the same amount of energy as the United States consumed in 1970 (China has more than four times the population of the U.S.). If China succeeds in nearly doubling its per capita consumption by 2030, as forecast by the U.S. Energy Information Administration, it will be the largest consumer of energy in the world, but it will still only have one-quarter of the per capita consumption of the United States. Someone's dream of a better future will prove to be untenable, and as competition grows over finite supplies, price volatility and open hostility will mark relations between those laying claim to the planet's remaining hydrocarbons. There just will not be enough to go around.

This is not idle speculation, but rather a looming crisis we need to begin to prepare for right now. The roles oil, natural gas and coal play in our civilization are complex, far-reaching and probably irreplaceable. A world marked by scarcity is something beyond the scope of this essay to imagine. But to give a sense of the scale of the crisis facing us, consider the role of oil and gas in producing food.

Global population is forecast to reach more than 9.2 billion people by 2050, from 6.7 billion at present.[22] The question of how we will feed all these people is crucial. The Green Revolution has been powered by hydrocarbons: diesel fuel, pesticides, herbicides and natural gas–based fertilizers. Before nitrogen from petroleum-based fertilizers was available, typical corn yields were 30 percent of what they are today.[23] Our modern agricultural system yields only one calorie of food energy for every ten calories of fossil fuel energy invested in producing it, a very poor EROI considering the ratio was 1:1 at the beginning of the 20th century, and it only gets worse when we factor in the energy costs of shipping, packaging and marketing the food.[24] But as inefficient as this system is, billions depend on it—and making fuel and fertilizer more expensive is not going to make this system any better, at least in the short or medium term.

World grain reserves are at the lowest levels since record keeping began in 1960, and at the lowest levels in the U.S. since 1948.[25] There have been food riots in many parts of the world due to rising prices.[26] This has been blamed on the diversion of food crops to biofuel crops, but the main reason is the rising cost of hydrocarbon inputs for food production. The looming issue of fossil fuel depletion will put ever more pressure on food production. As a result, there will be higher food prices and greater civil unrest, particularly within the large segment of the world's population that is living at close to subsistence levels. Peak oil and natural gas means peak food, without a rethink of our food production systems.

Now, consider that food production is just one of the countless fields in which hydrocarbons have become an indispensible aid, and the sheer scale of the crisis we face in the shadow of hydrocarbon depletion starts to become clear. The implication is that we must recognize the limitations of these fuels and rationally manage their future use if we are to avoid chaos.

The absolute first priority is to realize that renewable energy technologies, even if rolled out to their maximum, cannot hope to fill the energy-demand void left by hydrocarbons at our current business-as-usual levels of consumption (let alone forecasts). The void is just too large. Although the amount of solar energy arriving on the planet each day is vast, the conversion of this energy to phytomass, which in turn is converted to fossil energy in the form of oil, gas and coal, is a very inefficient process. Earth produces about 105 gigatonnes of carbon as phytomass each year from solar insolation and has stored phytomass of 500–600 gigatonnes of carbon. Given the preservation rate from phytomass to hydrocarbons of about 7000:1, the hydrocarbons burned each year amount to the hydrocarbons that could be accumulated in five hundred years of Earth's phytomass production, or eighty to one hundred times the amount that could be produced from the total phytomass stocks of the planet.[27] All of which is to say what we already know—we are burning precious stocks of hydrocarbon energy at rates far beyond Earth's ability to replace them.

And we are going to need this energy if we are going to manage the transition to a non-hydrocarbon energy infrastructure. For example, the construction of renewable alternatives, be they windmills, photovoltaic cells, tidal turbines, micro- and large-hydro equipment, solar-heating equipment or geothermal equipment, all require materials mined, processed, transported and manufactured by hydrocarbon energy. Given that most of these renewable resources produce only electricity, as opposed to energy-dense liquid fuels, it is not possible to build much of the infrastructure required to go beyond hydrocarbons using renewable energy.

Ask any renewable energy dealer what is the first thing to be done in designing a renewable energy system and he will tell you to reduce energy loads as much as possible first, before considering anything else. And that is what we must do as a society. A paradigm shift in our approach to energy consumption is required, with the realization that, in the future, energy supplies will be much more expensive and, in some cases, not available in the continuous supply mode that we have become so accustomed to. Risk management dictates that we reduce our vulnerability to energy-supply shortages and disruptions by first reducing our needs as much as possible and then by filling remaining requirements with an objective and pragmatic selection of available energy options—there are clearly no silver bullets, but there are many small incremental silver BB's that with luck and good management could add up to a longer term solution. European and Southeast Asian countries like Japan and South Korea can teach North America many things about reducing per capita consumption. These countries have a per capita energy consumption level of less than half that of the U.S. Excellent mass transit, much better rail systems and higher population densities are some of the things that have allowed reduced energy consumption in these countries, while still providing a First World standard of living. Germany and Denmark, among other countries, are world leaders in the application of renewable energy.

That said, hydrocarbons will of necessity continue to be an indispensible part of our energy future, and will provide a significant proportion of our energy for decades to come. Their finite nature must, however, be front and centre in planning our future energy-security strategy. After we implement the many opportunities to reduce consumption, we must use the hydrocarbons that have to be burned in the most efficient manner possible. We must recognize their unique one-time nature.

The apocalyptic visions of climate change portrayed by many climate scientists have provoked much-needed debate. Many mitigating options have been proposed: from carbon taxes and cap and trade to carbon capture and storage; from putting shields in space to filling the atmosphere with sulphate crystals to seeding the oceans with iron filings. These apocalyptic visions are for the most part based on the mistaken assumption that there are more than enough recoverable hydrocarbons to fry the planet several times over. It is also commonly assumed that maintaining growth without adverse effects is just a matter of converting to alternative forms of energy (notwithstanding the issues of scale and other limitations discussed earlier), burying the problem or somehow deflecting sunlight through uncontrolled experiments with the atmosphere and in space. We seem willing to try almost anything to cling to our business-as-usual paradigm. But we would be worse than fools if we were to squander the remainder of our inheritance of fossil energy in pursuit of this impossible dream.

Our belief or disbelief in the finite nature of hydrocarbons and the likely near-term peaking of hydrocarbon supplies will profoundly influence how we deal with the twin issues of energy sustainability and climate change. If we believe in a world of finite hydrocarbons, it will be necessary to implement strategies to reduce consumption through conservation and efficiency and to implement ways to reduce the consequences of supply disruptions. If we believe in a world of nearly infinite hydrocarbons, other strategies which increase the consumption of hydrocarbons may also be considered, including extra energy expenditures for carbon reduction through sequestration (about 30 percent of the energy in a typical coal plant is required to capture and compress carbon dioxide), launching solar shields into space and seeding the atmosphere with sulphate crystals. As a geoscientist who has spent the better part of four decades studying and quantifying hydrocarbon resources, I believe the nature of the evidence for hydrocarbon depletion and peak deliverability is overwhelming, and that we had best get on with planning our future in this particular context.

At the end of the day, the easiest and least costly way to

reduce greenhouse gases is to avoid burning the hydrocarbons in the first place. In other words, we need to implement ways to reduce consumption and our vulnerability to supply disruptions, and the good news is that there are many ways this can be done. This addresses both the Energy Sustainability Dilemma and climate change. It also preserves some of Earth's legacy of fossilized sunshine for future generations. Humankind has had a really good run over the past century, as a result of the exploitation of non-renewable energy and other natural resources. The party's coming to an end, and to avoid an extremely painful hangover we need to be aware of where we are at and where we must go. This will require an objective assessment of all of our options—nothing can be left out. We will likely need them all.

Mark Jaccard

PEAK OIL AND MARKET FEEDBACKS:
Chicken Little versus Dr. Pangloss

Mark Jaccard has been professor in the School of Resource and Environmental Management at Simon Fraser University, Vancouver, since 1986—interrupted from 1992 to 1997 while he served as Chair and CEO of the British Columbia Utilities Commission. Jaccard has served on the Intergovernmental Panel on Climate Change (1993 to 1996), for which he was awarded the Nobel Peace Prize in 2007, and the China Council for International Cooperation on Environment and Development (1996 to 2002). He has chaired several public inquiries, advised governments throughout the world and is a frequent media contributor. He is a member of Canada's National Roundtable on the Environment and the Economy, a research fellow at the C.D. Howe Institute and lead author for policy in the upcoming Global Energy Assessment. He has over ninety academic publications. His 2002 book, *The Cost of Climate Policy*, won the Policy Research Institute award for best policy book in Canada and was shortlisted for the Donner Prize. His 2005 book, *Sustainable Fossil Fuels*, won the Donner Prize for the best book on Canadian Public Policy. His 2007 book, *Hot Air: Meeting Canada's Climate Change Challenge*, is co-authored with Nic Rivers and Jeffrey Simpson of the *Globe and Mail*. Jaccard was named 2008 Academic of the Year by the association of British Columbia faculty members.

As portrayed by its most ardent promoters, the increasingly popular peak-oil hypothesis suggests that the sky is falling and that we need to be worrying about this a great deal. The earth's crust has a fixed supply of oil and we can estimate just how much based on our past experiences in finding and exploiting it. Indeed, since we know the amount of oil we have already extracted and the rate at which we currently exploit it, we can draw a line into the future and predict the date when our production of oil will peak. This simple calculation allows experts and non-experts alike to announce with confidence the time when we will start to run out of oil. Some pundits seem to enjoy providing the date with almost biblical precision, pinpointing that rapturous day when automobiles are abandoned by the roadside and suburbs deserted overnight. A scan of any popular bookstore confirms that there is a vibrant market for the peak-oil hypothesis and for depictions of the dramatic events that it will presumably trigger.

But is there any truth in it? Will the depletion of a non-renewable and essential resource like oil cause a cataclysmic shock to our economic system? Will the demise of oil happen rapidly? Will its depletion render obsolete, within a very short time, much of the road infrastructure, the suburban housing, the dispersed shopping malls, the stocks of trucks and cars, as well as all the industries that are dependent upon the long-distance shipment of goods?

While it is certainly not accurate to lump all economists together, many are nonetheless associated with a more optimistic view of how our economy will respond to the depletion of a key resource such as oil—a view that some compare to the unshakable optimism of Dr. Pangloss in Voltaire's famous book *Candide*. Pangloss's explanation for every mishap, no matter how troubling, is that all is well in the best of all possible worlds. The fundamental training of economists has taught them that economic systems are anything but linear, that they respond to real and perceived scarcity in ways that are likely to offset and perhaps avoid the cataclysmic outcome portrayed by devotees of the peak-oil hypothesis. In essence, economists note that if a good is valued, and if it is becoming scarcer, then its price will rise. When this happens in a market economy, several feedback mechanisms kick in. These price responses can slow down and perhaps even reverse the upward price trend. In other words, the rules of the game change when the price changes.

In the case of oil, we have recent historical evidence in support of the views of economists. In the not too distant past, we have had very high oil prices. Initially in 1973, and then more substantially in 1979–80, oil prices rose from three dollars per barrel to a peak of more than forty dollars per barrel. This was the price in 1980 dollars, but when converted after inflation into today's money, the initial price was seven dollars and the forty-dollar peak was ninety dollars. It is only in 2008 that oil prices—seemingly so astronomical—finally surpassed their peak of 1980. Figure 1 provides a long history of oil prices in current values. Because the figure lacks average prices for 2008, it does not show the lofty price of $140 reached temporarily during that year.

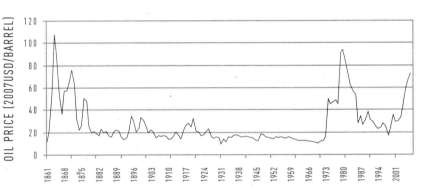

Figure 1. World oil prices from 1861 to 2007.

(Source: *British Petroleum Statistical Review of World Energy 2008*)

Obviously, a jump, in today's money, from seven to ninety dollars for a barrel of oil in just seven years is a huge price increase. What is interesting, however, is what happened during and after this increase. Just as most economists predicted, the market's feedback mechanisms kicked in—with a vengeance. The high prices triggered a range of supply-and-demand response.

On the supply side, the higher oil prices stimulated several responses from the oil and gas industry. It searched more intensively for oil, in anticipation of higher profits. It pursued technological innovations that would increase the amount of oil that could be extracted from a given site. It reassessed the magnitude of its reserves, as higher prices mean that more of the estimated reserve can be profitably exploited. In concert, these responses increased the available supplies of the oil resource.

Also on the supply side, the higher oil prices stimulated investment in alternatives to conventional oil. Some investors supported the development of fuels from unconventional oil like oil sands, heavy oil and oil shale. Others supported the development of synthetic oil substitutes from the other fossil fuels: coal and natural gas. Still others put their money in the development of biofuels, with the Brazilian government especially noteworthy for its support of ethanol from sugar cane as a key component of its energy self-sufficiency strategy.

On the demand side, business and consumers found ways to use less oil, either by improved efficiency of vehicles, furnaces, boilers and so on or by switching from oil-based products to energy substitutes like coal, natural gas, nuclear, hydro power, wind power and so on. By 1986, what had been a seller's market for oil transformed into a buyer's market. The price crashed. And it stayed below twenty-five dollars per barrel for most of the next fifteen years.

This appears to be a textbook case of the wonderful power of the price-feedback mechanism of the market. A seminal economic thinker like Adam Smith would be pleased with the manner in which the energy-use decisions of millions of businesses and households enabled the global oil market to establish a new equilibrium price, one that held more or less for the next decade and a half, with governments playing a minor role at best. No doubt, economists enjoy teaching the lessons from this particular event.

But what are the lessons really? Does the response to the oil-price shocks of the 1970s demonstrate the ability of the market to deal with the gradually increasing scarcity of a non-renewable resource? Or does it mislead us into thinking that the market will take care of resource depletion when in fact it will not? Was the event really just a foreshadowing—an ignored foreshadowing—of much more dramatic events yet to come?

Those focused on peak oil argue that while the rapid oil price increases of the 1970s were sometimes interpreted as a response to geological scarcity, they were actually the result of an artificial scarcity motivated by geopolitical factors. The rapid price increase in 1973 coincided with the oil embargo initiated by the major Arab oil-exporting countries in response to that year's Arab-Israeli War. The further price jump in 1979 coincided with the sudden cessation of oil production in Iran during its Islamic Revolution. This type of scarcity is resolved once the political situation changes. Arab countries soon ended their 1973 oil embargo. In the early 1980s, the new Iranian government re-established its oil exports, in part to finance its costly war with Iraq.

In other words, those concerned about peak oil claim that the oil price increases of recent years are not comparable to what we experienced in the 1970s. They argue that the high price of oil in 2008 was directly linked to physical realities. It is not that producing countries aren't selling us enough oil. It's that they cannot sell us any more. They are pumping as fast as they can, the argument goes, but relentlessly growing demand from developed and especially developing countries is soaking up the oil as fast as it comes out of the ground. They acknowledge that the oil market is, of course, still somewhat vulnerable to political developments. But they go on to argue that the price increases in 2007–2008 were not ephemeral, and instead signify a new recognition that the severe scarcity of oil is now a reality in determining its price.

This is a critical question. Depending on who is right, the future will indeed be very different. On the one hand, we have the Chicken Little perspective that the sky is falling, that we should quickly get ready for oil prices at two hundred, three hundred and even four hundred dollars per barrel, and that our economies, our industries, our cities and even our lifestyles must rapidly change. On the other hand, we have the optimistic perspective of Voltaire's Dr. Pangloss, that in this best of all possible worlds the economy's feedback mechanisms will kick in while our economies transition gradually towards the many alternatives to conventional oil, at energy costs that are not dramatically higher. From Dr. Pangloss's perspective, we will look back at this first decade of the twenty-first century as nothing more than a bump on the road of economic growth.

Which side is right? How much oil do we have left? What are the fossil and non-fossil fuel substitutes to oil and what supplies are available at what cost? What is accurate and what is inaccurate about the peak-oil depiction of the world? Given all these considerations, can we say anything about where the price of oil might be in ten, twenty or thirty years?

Is the Tank Really on Empty?

The first step in answering this question is to ask how much fossil energy is left in the ground. The short answer is "a lot."

In my 2005 book, *Sustainable Fossil Fuels*, I provide detailed estimates of the world's supply of oil, coal, natural gas and other forms of energy. A table from that book is reproduced here (see figure 2). This table shows my supply estimates for the three fossil fuels; my calculations are assembled from various official sources.[1] The table makes an important distinction between resources and reserves. While they may sound like the same thing, they are actually quite different. A "resource" is defined as the estimated natural occurrence of a particular form of energy—how much there is in the ground. A "reserve," on the other hand, is the subset of the resource that is available for current exploitation— that is, how much we're likely to get out of the ground at today's technology and prices. The size of the reserve depends on our state of knowledge of the resource's location and our technological capability to extract and process it at a reasonable cost, so it is difficult to set an exact figure. The definitions of "resource" and "reserve" differ from agency to agency, and there is a further distinction between "proven" reserves and "probable" reserves. To further complicate matters, different agencies use different ratios to convert physical quantities (tonnes, barrels, cubic metres) into energy values. But the numbers are all there, and they give a rough sense of how much energy is still in the ground. Since many of my numbers are from chapter 5 of the *World Energy Assessment* of 2000, I use its energy conversion ratios.[2]

The world coal resource is estimated to be over 7 trillion tonnes or 200,000 exajoules. World coal reserves are estimated to be one trillion tonnes or 21,000 exajoules. To get a sense of these magnitudes, keep in mind that humanity

Fossil Fuel	Production in 2000 (EJ)	Total Reserves (EJ)	Total Resource (EJ)	Reserve / Production in 2000 (years)	Resource / Production in 2000 (years)	Resource / Production with growth (years)
Coal	100	21,000	200,000	210	2,000	< 400
Oil (total)	163	11,000	32,000	67	196	< 150
Conventional		6,000	12,000			
Unconventional		5,000	20,000			
Natural Gas (total)	95	15,000	49,500	158	521	< 300
Conventional		5,500	16,500			
Unconventional		9,500	33,000			
Total Fossil Fuels	358	47,000	281,500	131	786	

Figure 2. World fossil fuel resource estimates.

(Source: M. Jaccard, *Sustainable Fossil Fuels*)

Notes:

• Unconventional natural gas does not include geopressurized gas and gas hydrates.

• My assumptions for the last column include the following: coal grows to its business as usual level of 650 exajoules in 2100 (a 1.9 percent annual rate) and at 0.5 percent thereafter; oil grows from 2000 at a 0.5 percent annual rate; and natural gas grows to its business as usual level of 160 exajoules in 2100 and continues at a 0.5 percent annual rate.

currently uses just over 400 exajoules per year in total (which is equivalent to about 2 million barrels of oil per day) when we include biomass and other renewables alongside the dominant fossil fuels. These coal reserves are concentrated in North America, the former Soviet Union, China, India, Australia and sub-Saharan Africa. (For simplicity, the table presents coal as a homogenous resource. In reality, differences in the grades of coal are important, with the highest-grade coal used as a feedstock and energy source in steel production while lesser grades are combusted for steam production in electricity generation plants.) Coal reserves are substantial compared to our current use rate, to say the least. If coal consumption continued at its current rate of one hundred exajoules per year, reserves might last 210 years, while the resource would last 2,000 years. These long time frames would decline of course if the exploitation rate were to increase or if not all of the resource were ever to become technologically or economically accessible. But it is clear that there is a lot of coal out there.

And the same is true for oil. Statistics of oil reserves and resources make a distinction between conventional oil and unconventional oil. Conventional oil is generally defined as crude oil which can be pumped to the surface without requiring induced changes to its viscosity (low-viscosity oil is more fluid). The cost of recovering crude oil varies with the location and characteristics of the reservoir. Some conventional oil is pooled in large, shallow reservoirs and flows easily to the surface. But some conventional oil is more difficult to exploit because of low pressure, the depth of the reservoir, the heaviness of the oil or the remoteness of the location—hence terms like deep oil, heavy oil, offshore oil, deep offshore oil and frontier oil. Technological developments have improved the ability to exploit these sources of conventional oil and have also increased the percentage that can be extracted from a given reservoir. A technique called "enhanced oil recovery" injects a gas or liquid into an oil reservoir to raise its pressure and change the oil's viscosity in order to improve the recovery factor; traditional recovery factors of 35 percent have been increased to more than 65 percent in some cases. This can make it difficult to determine just what is a resource and what is a reserve. But either way, the oil is there, under the ground.

"Unconventional oil" describes various substances from which crude oil or a synthetic variant can be produced. Oil sands (also called natural bitumen or tar sands) are the form of unconventional oil that gets the most headlines. They consist of loose-grained rock material bonded with bitumen, a molasses-like heavy oil. Synthetic oil is produced from oil sands by heating processes that can be applied in situ or after the physical removal of the material by strip mining. The choice of process depends on the depth of the deposit, production costs and environmental impacts. Western Canada has a significant portion of the world's resource, and has recently seen a major expansion in production. The potential scale of the resource is comparable to the conventional oil resource of Saudi Arabia. Extra-heavy oil is a tar-like substance requiring steam injection for recovery as synthetic crude oil. Venezuela has about half of the world's extra-heavy oil resource and is currently developing some of it. Oil shale is a fine-grained sedimentary rock that contains a waxy hydrocarbon material called kerogen. Large volumes of this rock must be mined and then subjected to heat processes in order to yield petroleum products. The U.S. has large deposits of oil shale but has only conducted pilot projects to process it. Estonia, however, has exploited its oil shale resource for electricity generation for decades.

Again relying primarily on the *World Energy Assessment*, figure 2 provides resource and reserve estimates for conventional and unconventional oil. The conventional oil reserve is estimated to be six thousand exajoules (that is, 150 gigatonnes), while the resource base is estimated at between ten thousand and sixteen thousand exajoules. Estimates of unconventional oil resources have even greater variability given the uncertainties about magnitude, future technological developments and ultimate exploitation costs. The *World Energy Assessment* estimates the unconventional oil reserve at 5,000 exajoules and the total unconventional resource at 20,000 exajoules. If global oil consumption continued at its current annual rate of 163 exajoules, currently estimated reserves would last sixty-seven years and the estimated resource 200 years. If oil production were instead to grow from 2000 onward at 0.5 percent per year, the estimated resource (conventional and unconventional) would last less than 150 years.

And natural gas is even more plentiful. As with oil, analysts distinguish between conventional and unconventional natural gas. Conventional gas is natural gas in pressurized sedimentary structures in association with oil or by itself. Even gas that is not found with oil is usually mixed with other gases, hydrocarbon liquids and water, thereby requiring processing prior to its long-distance transport to market by pipeline.

Unconventional natural gas is gas that cannot be exploited by conventional recovery techniques. Most analysts distinguish four types of unconventional gas. Coalbed methane is gas that can be extracted from hard-coal deposits, usually requiring fracturing of the coal. Tight-formation gas is gas trapped in low-permeability rocks that requires fracturing to be released. Geopressurized gas (also called ultra-deep gas) is gas dissolved in deep aquifers. Gas hydrates (also called clathrates) are deposits of frozen gas and ice located in polar permafrost and below sediments on the ocean floor.

Estimates of conventional natural gas reserves depend on assumptions about the economic potential for enhanced gas recovery; for example, on raising reservoir pressure to increase recovery from 60 percent to 70 percent. Assuming that some enhanced recovery is economic, the *World Energy Assessment* estimates global reserves of conventional natural gas to be 5,500 exajoules and the entire conventional resource to be 16,500 exajoules—of which perhaps 5,000 exajoules could be added to the conventional gas resource from enhanced recovery methods.

Again, as with oil, estimates of unconventional gas reserves and resources vary widely depending on some key assumptions, such as whether to include the huge but highly uncertain estimates of gas hydrates and geopressurized gas. The *World Energy Assessment* estimates that unconventional gas reserves are in the range of 9,500 exajoules while unconventional gas resources are in the range of 33,000 exajoules. These totals exclude the enormous quantities of gas hydrates, approximately 350,000 exajoules, and geopressurized gas, estimated at 600,000 exajoules. My exclusion of these latter two is questionable, however, because while humanity has not yet developed the capability to exploit these resources, some analysts believe that the technological means could be quickly developed at reasonable costs. But I leave them off the ledger to make my estimate more cautious.

Combining conventional and unconventional gas yields a total gas reserve of 15,000 exajoules and a total gas resource of 49,500 exajoules. If global natural gas consumption continued at its current annual rate of 95 exajoules, reserves would last 160 years and the resource as long as 520. Just to give a sense of the potential of some of the unconventional gas, if hydrates and geopressurized gas were included in the resource total, the resource could conceivably last thousands of years!

Since all three forms of fossil fuel can be converted into liquid and gaseous hydrocarbon forms of secondary energy as well as electricity and hydrogen, figure 2 combines the reserve and resource estimates into a single fossil fuel aggregate. The total of fossil fuel reserves is 47,000 exajoules and the total of fossil fuel resources is 280,000 exajoules. As an aggregate, fossil fuel reserves would last 130 years at current-use rates and resources almost 800 years. I have not projected an indefinite growth rate for fossil fuel consumption in aggregate, but from the individual resource projections in the table it is apparent that the total resource could support an annual growth rate of 0.5 percent for about three hundred years—again excluding the huge potential from gas hydrates and geopressurized gas.

Now, these estimates are highly uncertain, since they depend on the state of geological knowledge, the rate of technological advance and the price of energy. But even if these resource estimates are on the high side, it is nonetheless clear, from the sheer magnitudes involved, that humans have thus far used only a small portion of Earth's fossil fuel endowment. From a supply-security perspective at least, it appears that we can continue to use these resources for decades and even centuries—provided we do this without causing unacceptable disruption to Earth's climate and key biophysical systems.

We're not running out of fossil fuels just yet.

Peak-Oil Debates:
Is the Peak in Conventional Oil Production Predictable?

Those who warn us of the looming crisis of peak oil seem to be generally aware of Earth's huge fossil fuel endowment. And yet, when it comes to energy security, they seem to draw the opposite conclusion: that oil penury is already upon us and that oil prices are inevitably headed to ever higher levels. To back up this position, peak-oil champions refer to the pioneering analysis in the 1950s of the U.S. geologist M. King Hubbert. He postulated that discoveries of oil reserves in a given oil-bearing basin would trace a bell-shaped curve over time. At first discoveries would increase, and then they would hit a peak, after which they would decline as recovery efforts in the same basin faced diminishing returns. Oil production would have the same bell shape as the reserve curve, but it would lag behind it by about fifteen to twenty-five years, given the time it takes to incorporate new reserves into production. The right tail of the production curve signifies depletion of the reserves and eventually the cessation of production.

Using U.S. exploration data, Hubbert determined in 1956 that oil discoveries in the lower forty-eight U.S. states were reaching a peak and from this he predicted that peak production would occur sometime between 1965 and 1970.[3] As it turned out, his prediction was quite accurate, with peak production occurring in 1970. Figure 3 depicts the bell-shaped curve, showing the addition of oil reserves over time as a result of exploration effort and the physical limits of the resource.

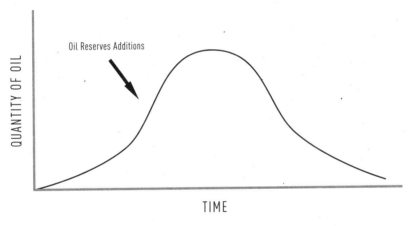

Figure 3. Hubbert's model of peak oil.

(Source: author)

Peak-oil advocates have extended Hubbert's model to the global scale. C. Campbell, J. Laherrère, R. Bentley, K. Deffeyes and other researchers have predicted that global oil discoveries have peaked and that global oil production will hit "Hubbert's peak" within a decade or two at most.[4] Similarly, some researchers predict that natural gas production will peak after 2020, and then follow a similar pattern of decline. These advocates of a global application of Hubbert's model base their analysis on a detailed review of the discovery and production rates in the world's currently active oil and gas fields, plus an assessment of the potential from less intensively explored sedimentary basins. While they acknowledge the backup potential from unconventional oil and gas, they suggest that these cannot be brought into production fast enough to offset the growing gap between declining production and increasing demand. They also claim that because these unconventional sources consume energy in the production process, their use will entail a declining productivity in our energy system, which will imply a similar decline in economic performance. As a result, they predict a crisis in oil and gas markets because of fuel scarcity, sustained high prices and global economic recession. Populist writers have played up the resulting crisis potential, with daunting book titles such as *Peak Oil, The End of Oil, Out of Gas* and *The Party's Over*. Imminent catastrophe has its attractions.

Again, to experts and non-experts alike, this hypothesis seems highly plausible. Oil is a non-renewable resource whose discovery and production must eventually decline. Nothing could be more obvious.

Enter the economists, for whom oil resources are not strictly limited. In the words of E. Zimmerman back in 1951, "Resources are *not*, they *become*."[5]

What Zimmerman meant by this is that the concept of a resource is much more fluid. When we only had technologies to extract oil from land-based oil wells, then oil under the seas was not a resource. When we developed the offshore technology, then undersea oil became a resource. In more recent decades, oil sands were not oil resources. Today, they most decidedly are. Perhaps the same development will occur with oil shale, which is rarely counted as an oil resource today but may be in future. Perhaps the same development will occur with a substantial part of the 30 to 60 percent of oil in a given reservoir that is considered today to be unrecoverable and is therefore excluded from our resource and reserve estimates.

This explains why economists have much more difficulty estimating something like the "peak" in oil discovery or production. The peak is at different times if different definitions of the oil resource are used—and the definition is indeed changing all the time. Recent studies by economists have given very different estimates for the peak of production depending on the associated estimates for: (1) production cost reductions caused by innovation, (2) reserve reassessments caused by price increases, and (3) capacity increases through more intensive exploitation of individual wells.

What Happens As We Deplete Conventional Oil Reserves?

Most of those who are concerned with peak oil do not exhibit any particular love for oil. In fact, one sometimes wonders if their certainty about its imminent disappearance is not, in part, the result of unconscious wishful thoughts. Some of the writers about peak oil seem to barely conceal their glee in describing the collapse of our modern economies and the forced return to an oil-free, pre-industrial, bucolic lifestyle. One gets the sense that they feel this would be good for us.

Perhaps this is where we are headed. And perhaps this would be good for us. But what is the likelihood of this scenario? What if we really were on a downward slope in the discovery and production of oil? What would happen? Would prices really rise inexorably as global output of conventional crude slid down the right side of Hubbert's curve?

To most economists, the probability of this happening is extremely low. Every commodity, no matter how essential, no matter how coveted, has factors that limit its price. This limit might be determined by the cost of producing alternatives to the commodity—the price of its substitutes. At the extreme, it might be determined by the limits of our ability to spend—our budget constraints. But there are limits.

The peak-oil literature—consistent with Hubbert's analysis—tends to rely on data for conventional oil. Earth's endowment of unconventional oil is all too often ignored. Yet humanity would be nowhere near the possible production peak of oil if the curve were to include enhanced oil recovery, oil sands, extra-heavy oil and oil shale. All of these are already being exploited to some degree, but their production could be scaled up dramatically over the next decade or two. This has already started with enhanced oil recovery. And exploitation of the oil sands in Canada has taken off from a modest level in the mid-1990s, to reach more than one million barrels per day in 2008 with expectations to hit four or five million barrels by 2020.[6]

Even adding unconventional oil to the curve does not do justice to the full-supply response that would be triggered by higher oil prices. Conventional oil and unconventional oil are used to produce refined petroleum products. However, these refined products can also be produced from natural gas and coal. Today in Qatar, natural gas is used to produce synthetic gasoline and other liquid fuels. In South Africa, coal is used to produce synthetic gasoline, providing more than 40 percent of the fuel for vehicles in that country. As figure 2, discussed earlier, shows, Earth's endowment of coal is huge and the stock of natural gas, which is also huge, could be even larger if humans eventually access deep geopressurized gas and gas hydrates. But there has been very little investment in these alternatives, since for a long time the price of conventional oil remained below twenty-five dollars a barrel.

Now, with the oil price higher, investment would be expected to flow into these alternatives, since their costs of production are likely to be below one hundred dollars per barrel and perhaps even lower according to virtually every energy analyst involved in estimating the costs of our energy alternatives.[7] Figure 4, a graph from A. Brandt and A. Farrell, is a recent update of the energy-supply graphs that are frequently produced by energy researchers. The horizontal axis of the graph shows quantities of oil that could be produced from the conventional and alternative fossil fuel sources listed above (in trillions of barrels). The vertical axis shows the costs of using these sources to produce a product equivalent to conventional oil. Eor is the abbreviation for "enhanced oil recovery," GTL is the abbreviation for "natural gas to liquids," and CTL is the abbreviation for "coal to liquids." The vertical height associated with each fossil fuel resource reflects the range of cost estimates. Oil shale production costs have great uncertainty and will vary significantly depending on the accessibility of the resource in a given location—hence the range from twenty-five to ninety dollars per barrel. The lighter shading moving to the right for each resource signifies increasing uncertainty about its magnitude. At least two trillion barrels

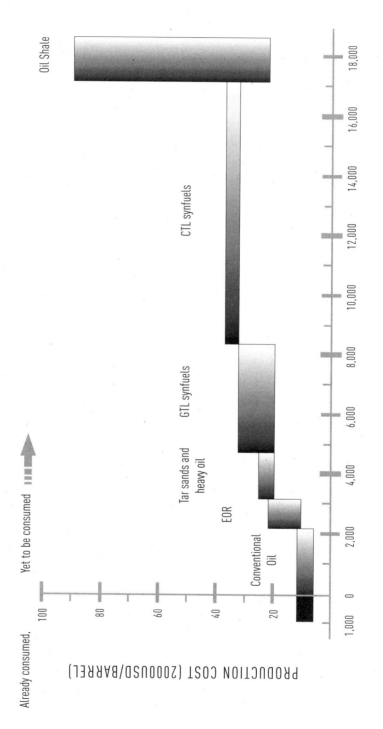

Figure 4. Conventional oil and its fossil fuel substitutes (in billions of barrels). (source: A. Brandt and A. Farrell, University of California Energy Institute, working paper, 2008)

of oil could be produced from coal (CTL synfuels), but the final figure could be as much as ten trillion. Finally, the solid bar to the left of the vertical axis shows that humans have so far consumed about one trillion barrels of conventional oil, most of it at a cost in the five-to-fifteen-dollars-per-barrel range.

Compared to the peak-oil diagram (see figure 3), this graph tells a very different story. The peak-oil curve gives the impression that we are on the brink of a steep downward slope. It leaves the question of what might then happen to the viewer's imagination. The possibility of dire consequences looms large.

In contrast, the fossil fuel supply graph provides an indication of what is likely to happen over the long term if market mechanisms are allowed to work. The rising price of oil, caused by a tightening market for conventional oil, will drive profit-seeking investment into these alternatives, all of which have already been exploited to some extent in commercial-scale operations for a diversity of market and non-market reasons. (South Africa developed its coal-to-liquids program in response to anti-apartheid oil embargoes decades ago.) Like virtually all other estimates by energy researchers, these alternatives to conventional oil appear to be profitable as long as the price of oil remains above forty dollars. Since the price has only been above forty dollars consistently for the last few years, it is no wonder that these alternatives to conventional oil today represent only about 3 percent of global oil production. But this percentage is growing every year and will accelerate if high oil prices hold for an extended period of time.

And other fossil fuels are not the only possible substitute for oil. Today, Brazil supplies more than half of the fuels consumed by its vehicles with ethanol produced from sugar cane. The U.S. government subsidizes the production of corn-based ethanol by its farmers. There are serious questions about the viability of this approach, given the relative inefficiency of producing ethanol from corn and the potential repercussions for world food prices. Nonetheless, production of liquid fuels from biomass is sure to expand over the coming decades, probably with a greater reliance on cellulosic feedstocks like sugar cane and wood, given the competition between land for food and land for fuel.

Finally, since a major demand for oil is the production of fuel for vehicles, it is important to note that vehicles can be powered by electricity and hydrogen, both of which can be produced without recourse to oil, or any fossil fuel for that matter. Battery-electric and plug-in hybrid vehicles could be powered by electricity that is produced from nuclear, large hydro power, biomass and other renewables. Hydrogen could be produced from electricity or biomass. The cost of these alternatives presents yet another long-run upper constraint on the price of oil, provided that they can be delivered at significant scale to affect the market price.

But even if it is true that there are many alternatives to conventional oil, it might still be the case that the transition will be very difficult, involving great costs that will lower our standard of living. In this regard, some people claim that we must especially pay attention to the amount of energy required to obtain a given amount of useful energy. Figure 5 explains the concept of "energy return on investment."[8] Energy return on investment (EROI) is the amount of useful energy provided to the economy (E_{net} in the figure) divided by the amount of energy self-consumed (E_{self}) and the amount purchased ($E_{purchased}$) by a given energy sector in production and delivery of the energy product. Conventional oil has a high return on energy investment, but most of its alternatives have a significantly lower return. Will this mean that our economy will be in big trouble, forced to exploit greater and greater quantities of energy just to provide the same level of energy services over the coming years as we deplete our conventional oil stocks?

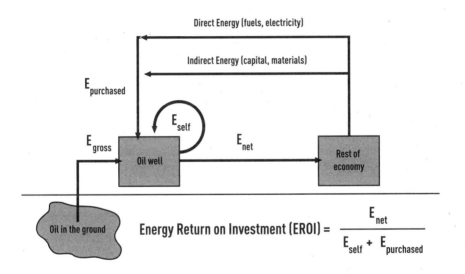

Figure 5. Energy return on investment.
(Source: Cleveland and Kaufman, *Encyclopedia of Earth*, 2008)

Once again, the answer is not that simple, at least not to economists. For economists, this calculation suggests that all energy provided by the energy system is equal, as long as we measure it in some common unit, like joules. But in the real world, if we converted energy prices into their per joule equivalencies, we would see that we value different energy products differently. For instance, sometimes it might make sense to use a lot of energy to produce an energy commodity that we value highly, like electricity. The value of joules in the form of electricity might be so great to us that the high energy input makes sense. In fact, what is important is not the amount of energy input to make electricity, but rather the capital and environmental costs.

Another example is provided by the oil sands of western Canada. Currently, natural gas is frequently used to upgrade the oil sands into synthetic crude oil, providing energy for the heat in the production process and for hydrogen to add to the carbon-heavy bitumen. When this natural gas is added to all of the other energy that is self-consumed by the oil sands industry—in extraction, processing and transport—the energy return on investment for oil sands is substantially lower than that for conventional oil. But this need not be the case. If the natural gas were more highly valued, then the oil sands industry would find an alternative source of energy, and the obvious alternative is to use more of the heavy fuel outputs of the bitumen itself to produce the heat that is needed to extract oil from the bitumen and to extract hydrogen from water in order to increase the hydrogen content. If the industry does this, the purchased energy will decrease, but the self-consumption of energy will increase. And the only limit on self-consumption is the cost of using more of the resource.

Thus, if oil sands exist free in nature, and if humanity can develop ways to extract this resource at reasonable capital cost without depleting natural water flows or negatively altering the value of the land (either by rehabilitating open-pit mines or by using less harmful in situ extraction of oil from the oil sands), then it does not really matter if the energy return on investment is lower than for conventional oil. All that matters is cost. Indeed, the previous figure that showed the cost of alternatives to conventional oil already included these extra capital costs. For these alternatives to be comparable to conventional oil in terms of environmental impact, however, their capital costs will probably be a bit higher than the figure suggests. Even so, this would mean that the costs of alternatives to conventional oil are well below a hundred dollars per barrel, for a long time to come.

The implication for the idea that we are running on empty is clear: even if we are running out of conventional oil, we are a long way from running out of fossil fuel energy.

Why Such High Oil Prices in 2008?

If markets work properly, one would expect this long list of reasonable-cost alternatives to keep the price of oil below one hundred dollars per barrel and indeed lower even than fifty dollars. Why, then, did the price climb so far above one hundred dollars in 2008? Are markets not working? Are the peak-oil alarmists correct in their Chicken Little forecasts of two hundred and three hundred dollars per barrel? What is going on?

In Dr. Pangloss's best of all possible worlds, markets would operate smoothly. The gradual depletion of conventional, low-cost oil would cause markets to tighten. This would shift the oil price slowly upward and investors would see that the price was not likely to fall back to its original level. On this basis, investments would steadily shift to alternative sources for producing the refined petroleum products we currently get from oil. Prices might edge upward, but only gradually as we shifted towards somewhat more expensive alternatives to conventional oil.

In other words, as one commodity (say, conventional oil) gets more expensive, the commodity that replaces it (say, unconventional oil) comes online. Because the cost of producing this next generation of fossil fuels is only slightly higher than the cost of what it replaces, the underlying cost of the gasoline it is refined to produce stays more or less the same. It may edge up gradually, but it will not spike. This gasoline-supply curve is illustrated in figure 6.

The graph shows the relationship between price and cumulative production of gasoline. The dotted lines depict the full underlying costs of producing and delivering gasoline, first from conventional crude oil, then from unconventional oil, then coal, and then perhaps a non–fossil fuel source such as biomass. If quantity A on the horizontal axis indicates cumulative demand at the corresponding price on the vertical axis, conventional oil supplies are sufficient. Over time, as we burn more and consumption drives cumulative demand to quantity B, gasoline suppliers must shift to unconventional oil. If unconventional oil is more expensive to extract and then refine into gasoline, consumers must pay more. If it costs about the same, the gasoline cost line is essentially flat and prices would be unaffected by the need to acquire supplies from this alternative. A similar logic applies to the supply and cost of producing gasoline from coal. Because of high uncertainty about future land-use conflicts, the graph shows two possible curves for the cost of gasoline (or, in this case, a substitute like ethanol) from biofuel: the cost could be relatively flat or rise steeply, depending on the value of alternative uses of land for agriculture or for the preservation of biodiversity.

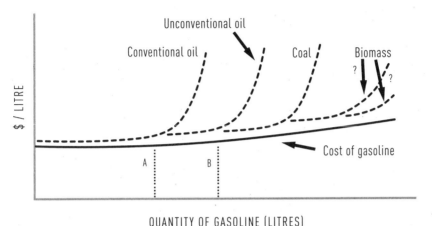

QUANTITY OF GASOLINE (LITRES)

Figure 6. Gasoline-supply curve.

(Source: Jaccard, *Sustainable Fossil Fuels*)

What will this mean at the pumps? If the cost of production of gasoline from the alternatives to conventional oil is not substantially higher in the long run, consumers will be unconcerned and probably unaware that substitution is occurring. All they may experience is a gradual increase in the market price for what appears to be the same product—gasoline to fuel their commute. This upward trend will likely be imperceptible in the face of the usual short-run price fluctuations inherent in any commodity, a normal consequence of the difficulty for markets of perfectly balancing supply and demand at any given time. In this scenario, peak oil becomes a meaningless concept, of no consequence to consumers.

Unfortunately for consumers, it does not always seem like we are living in Dr. Pangloss's best of all possible worlds.

Price changes for a critical commodity like oil are anything but smooth. In figure 1 we saw some of the major fluctuations in the price of oil over the decades. A closer look at it reveals that oil-price volatility is a common occurrence. These price spikes are mostly short-lived. But prices can also stay at high levels for much longer periods. To understand why, it is important to distinguish between the two key factors driving the demand for oil: its price and our incomes.

Energy economists refer to this relationship as the "double-elasticity function." By elasticity, they mean the change of oil demand resulting from a change in either oil's price or economic output (which determines our incomes). In other words, the more expensive oil gets, the less of it we will burn, and the more money we have, the more of it we will burn. The price elasticity of oil tells us how the demand for oil might change in response to a change in its price, with income and all other factors held constant (factors such as weather, the price of oil's substitutes or regulations on fuel efficiency). The income elasticity of oil tells us how the demand for oil might increase as our income increases, with price and all other factors held constant.

Energy economists make an important distinction between short- and long-run elasticities. The short-run price elasticity of oil might be quite modest. Even when the oil price rises significantly, consumers cannot instantaneously change their cars and furnaces to more efficient models or switch to alter-native fuels. So demand can only change a little (perhaps by driving less—a behavioural change) in the short run. But over time, the response is greater, meaning that the long-run price elasticity is higher. A sustained high price would motivate companies and households to gradually switch to more efficient devices or alternative fuels. So if a price stays high long enough, the impact of the price change increases.

In contrast to price elasticity, income elasticity indicates how the demand for oil might change if income is rising and the price of oil is not changing at all. Ever since the invention of the steam turbine and the internal combustion engine a century ago, the demand for oil has been strongly correlated with economic growth, an aggregate measure of income. Figure 7 shows how global oil production and gross world product have both grown over time. It is pretty clear that the more money we make, the more oil we consume.

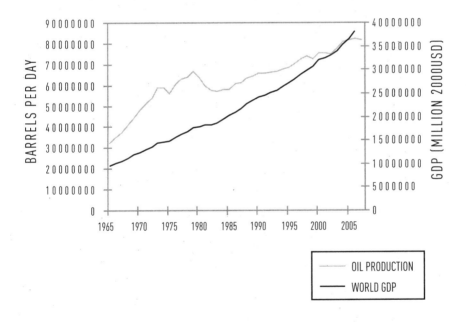

Figure 7. World oil production and gross world product.

(Source: World Bank, *World Development Indicators*, 2008, and A. Maddison, *The World Economy*, OECD, 2003)

But the link between a form of energy and economic output need not be inevitable. In the 1900s, economic output and coal production were closely linked. But as oil and natural gas appeared on the scene, the link of coal to economic activity weakened. The coal intensity of the economy—the amount of coal it took to make a dollar of output—dropped substantially. Today, the oil intensity of the economy is also dropping. How far and how fast depends on when and how fast we settle on something else to satisfy our energy needs. If oil is expensive in the long run, the line in figure 8, which illustrates the oil intensity of gross world product, may decline dramatically, as substitutes to conventional oil become relatively more affordable. But if the substitutes are mostly conventional oil, and if we redefine these substitutes as "oil," then the line might sustain its gradual slope.

But that is the big "if." Yes, there is a substantial probability that the price of oil will sustain its recent retreat below one

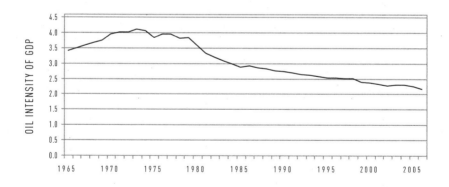

Figure 8. Oil intensity of gross world product.

(Source: World Bank, *World Development Indicators*, 2008, and A. Maddison, *The World Economy*, OECD, 2003)

hundred dollars per barrel; over time, commodity prices tend to find their way to the long-run cost of production. Still, there are also energy economists who believe that the price of oil will be well above one hundred dollars for many years to come, perhaps permanently. They note that the high prices of the last few years are the result of a very tight market rather than of conflicts in the Middle East or tensions with the government of Venezuela. The demand for oil has been rising steadily, driven in large part by the rising incomes in China, India and other developing countries experiencing rapid economic growth. In other words, income elasticity in developing countries (they are burning more oil because they have more money) is putting a great upward pressure on the demand for oil, and this is happening even though the price has risen. Income elasticity may dominate price elasticity. Of course, the impact of the oil-price elasticity (reducing use with higher prices) will increase with time, which means that the line in figure 7 will take on a steeper downward slope, similar to the period 1978–1985, or perhaps even steeper. Although it is too early to say definitively, initial data suggest the oil intensity of global economic activity is indeed starting to fall more rapidly in response to the price increases of the last few years—just as most economists have predicted.

This is on the demand side, where oil price and income elasticities operate. There is also, though, a supply price elasticity. Just as consumers figure out how to burn less, suppliers become innovative and offer alternatives as rising prices make these more affordable. This elasticity indicates the response of alternatives to conventional oil as its price rises. Earlier, figure 4 showed that there is a lot of oil potentially available at prices well below one hundred dollars per barrel. But this potential supply response is certainly not instantaneous—the short-run elasticity will be much smaller than the long-run elasticity. In other words, some years must pass before investors are convinced that prices will stay high. Then they start to make investments, but these investments can take years or even a decade to come to fruition. Meanwhile, the oil market stays tight and the price stays high, especially if economic growth in developing countries is offsetting the demand-price response.

Finally, these lags in the demand-side and supply-side responses to high oil prices can also induce speculation about just how tight the market might become and therefore just how high the price of oil might go. Some people are willing to pay very high prices for oil. In fact, many people will still drive their cars even when the price of gasoline is dramatically higher (that is, their demand is highly inelastic). Thus, a tight market induces speculators to buy and sell oil in the hope of profiting from quick changes in its price. The weight of speculator funds can thus cause the price to rise even higher during times of perceived scarcity. This phenomenon helps to explain the 2008 oil prices that rose above one hundred dollars per barrel. It adds yet another dimension to the factors determining the price of oil on a given day. As one U.S. politician apparently said, "It's startling to think of oil prices being set by young guys on the floor of the exchange screaming at each other on the basis of rumours."

Whither the Price of Oil?

Thus far, I have conveniently avoided saying where I think the price of oil will be in ten, twenty or thirty years. This is standard practice for economists. Say nothing or, if you must make a forecast, make sure to include a statement that begins, "On the other hand . . ."

Actually, in this chapter I have already made a forecast. I said that the long-run average price of oil will not diverge greatly from the long-run cost of production of conventional oil and its substitutes. Markets can get tight. In the short and even medium term, prices can rise dramatically, diverging greatly from production cost. Prices can even stay high for quite a long time because both the short-run price elasticity of alternative supply and the short-run price elasticity of demand are quite feeble—there are limited alternatives in the short run, so buyers will pay almost anything to keep consuming. But the long-run elasticities are likely to be quite substantial, especially given what we know about the costs of alternative supplies. These factors will make it very difficult to sustain a tight market indefinitely. The price of oil could still go higher. But eventually it will come back down towards the cost of production of oil and its substitutes.

On the other hand (there it is!), there is a scenario in which economic growth throughout the world and especially in developing countries might continue to overwhelm the price-elasticity effects on supply and demand, such that prices stay well above production costs for a decade or more. Quite frankly, I put a pretty low probability on this outcome. It is possible. But it is not likely.

And there is another factor that supports my prediction about price equating to production cost on average over the long run. The reader might wonder why I have not mentioned it yet. That factor is policies to address the climate change risk. I am convinced that humanity must impose a charge on the emissions of carbon and other greenhouse gases, and that this charge must grow gradually over the coming decades to be a very significant cost factor in determining energy supply- and-demand investment decisions. I call it a carbon price to remain neutral about whether this charge results from carbon taxes or economy-wide emissions caps with tradable permits.

Whatever pricing policy is used, we must quickly put a global price on carbon and other greenhouse gas emissions. We should have implemented this almost twenty years ago, at the time Scandinavian countries started in this direction on their own. This is a normative statement. I am not predicting that humanity will put this price on emissions. I am simply say- ing that it ought to and I hope it does.

If we do succeed in getting a price on emissions, there will be significant repercussions for the oil market, not to mention the other fossil fuels. This is because the rising price attached to carbon will form a wedge between the cost of producing refined petroleum products and their market price. The mar- ket price to consumers will keep rising in step with the carbon price, even though this upward pressure will reduce production (from what it otherwise would be) and therefore cause a coun- tervailing downward pressure on the prices that oil producers can charge. In other words, the carbon price will slow down the consumption of oil and therefore the progression from lower-cost oil to higher-cost oil and its higher-cost substitutes.

But does this mean that the demand for oil and other fossil fuels will fall precipitously in step with the rising car- bon charge? I think not. As I explain in *Sustainable Fossil Fuels*, the answer to this question depends on the relative advantages and disadvantages of the alternatives for providing humanity with low- and zero-emission energy. In some cases, nuclear power will be favoured. In others, it will be various forms of renewables. But in many cases, the conversion of fossil fuels into electricity, hydrogen and low-carbon fuels will be an attractive option as long as we capture and safely store most of the carbon. It will be more costly. But all of the options for low- and zero-emission energy are more costly. If we make an effective effort over the next fifty years, then it is likely that the cost of energy services will be 25 to 50 percent higher than today. In a rich country, this means that energy costs might be 9 percent of a typical family's budget instead of today's 6 percent. I think our children and their children, the ones who will be paying this cost, will be happy with the choice we've made.

In other words, Chicken Little and Dr. Pangloss might each be partly right. The future will look different from the world we live in now. But though some difficult choices lie ahead, and higher costs may make us doubt we live in the best of all possible worlds, there is no reason to think the sky is about to fall because of peak oil.

Jeff Rubin
DEMAND SHIFT

Jeff Rubin is chief economist and chief strategist of CIBC World Markets. Ranked one of the top economists in Canadian financial markets over the last decade, he is respected worldwide for his comments on global energy markets, the world economy and financial markets.

Why did it take so long for soaring oil prices to once again deep-six the world economy? For over the better part of this decade, growth in world oil demand accelerated despite as much as a 600 percent rise in oil prices. That demand shift flew in the face of one of the most fundamental axioms of economics, namely the downward-sloping demand curve. Somehow Adam Smith's invisible hand has been misled or misguided into consuming more oil even though its price has been going through the roof.

Since the millennium, world oil prices had increased seven-fold from $20 per barrel to over $140 per barrel by the first half of 2008 before eventually succumbing to the very global recession they had caused. That move surpasses even the enormous price spikes of the two OPEC (Organization of Petroleum Exporting Countries) oil shocks that brought the world economy to its knees. Back then, of course, global demand buckled almost immediately, and oil prices quickly came crashing back to the ground, just as any reasonable economist would predict. But this time around, it took almost a decade of price hikes, and a world financial crisis to boot, to bring prices down, and even then only to levels that as recently as four years ago would have marked an all-time high.

As it turns out, there is nothing wrong with the economist's theory of demand. But there is plenty wrong with how that theory of demand is routinely applied to world energy markets. The answer to the riddle of rising consumption in a market of rising prices lies in a regional analysis of where oil demand has been coming from and, just as importantly, where it has *not* been coming from. Regional differences aren't just important; they are critical to understanding what is driving global oil consumption numbers.

Beneath the surface of healthy growth in world crude demand over the last seven years are two very divergent trends. While one part of the world is attempting to decarbonize its economies, the rest of the world is burning hydrocarbons at a record pace. The global average is simply the netting out of two opposite forces. In other words, a few of us are burning less oil and a lot of us are burning more. A lot more.

The demand for oil today is weakest where it's historically been the strongest. In the past, economists would have looked at North America and western Europe to gauge the pace of world demand; this decade, it is the bustling energy demand from developing countries that has driven the global pace. In many of these countries, it's literally been only a couple of decades since bicycles and rickshaws outnumbered motor vehicles on the road. Yet these are the very countries that today drive all the growth in world oil demand. At the same time, the demand for oil in the mature OECD (Organisation for Economic Co-operation and Development) countries has already peaked and will fall steadily even after their economic recessions are long over.

OECD oil consumption had already posted two back-to-back yearly declines before the 2008 recession hit, a drop unprecedented since the fallout from the OPEC oil shocks nearly three decades ago. Even in the U.S. economy, by far the largest oil- and energy-consuming economy in the world, crude consumption had not risen since 2005. Declines have already occurred in western Europe and Japan. So the economists are partly right—high prices have killed demand. But it has simply come back to life on the other side of the planet.

Whereas the OECD accounted for almost three-quarters of world oil consumption in 1990, today it accounts for barely more than half. At current growth rates, oil consumption outside of the OECD will exceed the group's consumption within five years—heralding a new era in global energy markets.

The rich countries are not setting the rules on either the demand *or* the supply side of the equation anymore. Whereas world oil demand grew at just under 1 percent during the twenty years following the second OPEC oil shock, that rate has been closer to 2 percent between 2000 and 2007, even as prices have been rising spectacularly. The logic of economics says that it is impossible. But in oil markets, the logic of economics gets turned on its head.

If you want to find out what has driven all that demand growth for oil in the developing world, start by looking at the car. Roughly 90 percent of every new barrel of oil consumed in the world goes for transport fuel like gasoline and diesel. Look at where car sales are booming and chances are you will see that oil sales are booming there as well.

North America may be the land of the car, but that's very much a rear-view mirror on where car sales have grown this decade. While car sales in mature markets like the U.S. and western Europe crawl at 2–3 percent per year in the best of times, they grow ten to twenty times that pace in the so-called BRIC countries (Brazil, Russia, India and China). China's car market is already the second-largest in the world, and it's only a matter of time before it overshadows the U.S. market.

China's auto sales surged by 20 percent in 2007 while car sales in Brazil and Russia were up an even more striking 30 and 60 percent respectively. While the global recession has since quashed these gains, the long-run prognosis looks promising as the cost of car ownership becomes increasingly affordable with the advent of cars from the likes of Tata in India, which will soon produce a model that will sell for as little as $2,500. Everyone, from Nissan to Volkswagen to Ford, is joining Tata in India and Chery in China in developing cars with sticker prices under $10,000 that cater to the demand of first-time car buyers in developing markets.

Not many people in these countries can afford to burn diesel and gasoline in expensive cars. But they are lining up to burn fuel now that cars are cheap. It is the savings necessary to buy a car, not the price of gasoline, that is the greatest obstacle to driving in those countries. And just as car prices are falling there, incomes are rising. The world's roads and oil markets just got a lot more crowded.

Car ownership must no doubt be a liberating experience for the citizens of the developing world, offering many comforts and conveniences. But, at the same time, it gives each of those households a straw to start sucking at what are already rapidly depleting world oil reserves. Booming car sales will soon see emerging economies account for three-quarters of the new vehicles on the road over the next three decades. Just as oil consumption outside of the OECD will soon surpass consumption within the OECD, the same is about to happen for car ownership. Within five years, the majority of the world's auto fleet won't be driving on the autobahns of Germany or the interstate highways of America but on the crowded roads of the developing world.

In a world of no or little oil-supply growth, driving becomes a zero-sum game. Rising gasoline consumption in emerging markets must require an offset in falling consumption somewhere else. If not, there simply won't be enough gasoline to accommodate those new drivers in Russia, India, China and Brazil while keeping existing drivers on the road in North America, Europe and Japan. In other words, someone is going to have to leave their car in the driveway.

And chances are that person lives in a market where drivers pay the most for gasoline, and where driving behaviour is the most sensitive to changes in gasoline prices. The bad news is that those markets are all found in the OECD. For every new Tata or Chery owner in the developing world, someone is going to have to come off the road in the OECD. Which brings us to the second reason why oil demand has been so strong outside of the OECD, namely, the price of oil itself.

In 2007, OPEC, together with two non-member oil-producing countries, Russia and Mexico, collectively consumed 13 million barrels of oil per day. To put that number in perspective, it is almost *twice* China's 7-million-per-day intake and some 1,000,000 barrels per day more than the entire oil consumption of all of western Europe that year. This group now constitutes the second-largest oil market in the world, second only to the American market. That is another way of saying that the oil-importing countries are now competing for supply with the exporters they depend on. It is not a competition they have much chance of winning.

Unlike the American market, where demand has not grown in recent years, demand growth in oil-producing countries ranks among the highest anywhere in the world. Oil consumption in OPEC countries from 2004 to 2007 rose at an average annual rate of 5 percent, well over double the world's pace. While the rapid growth in oil consumption among oil-producing countries has largely taken place off most analysts' radar screens, its contribution to world consumption has been every bit as great as the much-noted contribution from China.

What makes oil-producing countries so thirsty for their own product? The answer lies not with the cost of buying a car, as in the BRIC countries, but with the cost of filling its tank. From Caracas, Venezuela, to Tehran, Iran, the key driver behind turbocharged gasoline demand is the price of oil itself. Triple-digit oil prices may have meant four-dollars-per-gallon gasoline prices for American motorists, but you could always fill up at the Jed Clampett price of twenty-five cents per gallon at your local gas station in Hugo Chávez's Venezuela. And motorists are paying only marginally more when they fill up their tanks in Saudi Arabia or Iran, where gasoline sells in the forty-to-fifty-cents-per-gallon range.

If American or European motorists could fill up at those prices or anything even remotely close to those prices, OECD oil consumption would not have fallen for the previous three years. On the contrary—it would have spiked, and the sports-utility-vehicle market would still be thriving. But the reality of world oil supply dictates that OECD motorists will never see those prices again while drivers in OPEC countries will continue to see those prices, no matter how expensive oil ultimately trades on world markets.

OPEC is by no means a homogenous group; it's divided by geographic, cultural, historic and religious fault lines. Yet in countries as disparate as Venezuela, Saudi Arabia and Iran there has emerged a common domestic political consensus that citizens of those countries should have an unfettered right to consume as much massively subsidized oil as they can. Governments that have tried to challenge that right by raising domestic gasoline prices have quickly felt the people's wrath.

Take the summer of 2007 in Iran for example. The Iranian government attempted to put a collar on runaway fuel-demand growth by raising the price of gasoline. The announcement quickly sparked riots in Tehran as irate motorists took to the streets and even torched a number of gas stations. Fearing a replay of riots over previous price-hike attempts that saw an Iranian prime minister assassinated in 1964, the Amadinajhad government prudently backed down on its scheduled price hikes and imposed a gasoline voucher system to deal with the growing problem of fuel shortages.

The notion that Iran, the world's fourth-largest exporter of crude oil, should be plagued by domestic fuel shortages may seem incongruous to Western motorists. But fuel shortages are nothing new to Iranians. While Iran may be one of the world's largest crude producers, it is, at the same time, one of the world's largest importers of gasoline. Years of 5-percent-plus demand growth have long surpassed the limited capacity of Iran's aging and increasingly dilapidated refineries.

The Iranian government currently spends billions of dollars a year in domestic fuel subsidies, at the expense of huge operating losses for the state-owned National Iranian Oil Company. Those subsidies in turn drive explosive fuel-demand growth while discouraging investment in new refinery capacity. So great has the mismatch between gasoline-demand growth and refinery capacity become that Iran has turned into one of the world's largest importers of gasoline. American motorists who experienced long lineups at the pumps during the Iranian oil boycott of America may now find it amusing that the Iranian government is worried about supply disruptions from politically hostile countries because of its growing dependence on gasoline imports.

It's not much different elsewhere in OPEC. In fact, when it comes to domestic fuel subsidies, no one can match the generosity of President Chávez's Venezuela, where massively subsidized oil is not only the foundation for the country's social and economic policy but plays an increasingly important role in Venezuela's foreign policy. Not only does Comrade Chávez provide Venezuelan drivers with the biggest fuel subsidy in the world, he also subsidizes drivers in politically allied states like Cuba and Nicaragua, and has forged a deal with Spain that will see the European country provide Venezuela with medicine, building materials and new technology in exchange for discount-priced oil.

So great is the popular demand for fuel subsidies that, perversely, in many OPEC countries, higher world oil prices actually raise, not lower, their own oil consumption, in total defiance of conventional economic logic. The higher world oil prices go, the more export earnings there are to chase twenty-five-to-fifty-cents-per-gallon gasoline sold at the domestic pumps. That is, higher world prices will mean not only more money coming in to fuel an economic boom that, in turn, will lead to higher car sales and energy demand—they will also put more money in the coffers of governments that subsidize fuel prices, further opening the throttle on the demand for gasoline and diesel. An economist might say that income elasticities for oil are very high—the more money people in these countries have, the more oil they burn. But *price* elasticity never comes into play, because these motorists don't have to respond to rising world prices by cutting back. They just burn more.

And the more oil consumption grows in countries like Venezuela, Iran and Saudi Arabia, the less they have to export to the rest of the world, causing oil prices to rise even further. If you are Hugo Chávez, it is a virtuous circle, where self-indulgence leads to self-enrichment. But if you are a North American motorist, it's a cycle that will take world oil prices so high that you will eventually be taken off the road.

OPEC is not only extravagant in its gasoline consumption; it burns oil in other ways that few other countries could afford. While the notion of burning crude oil to produce electricity may seem absurd in the face of recent oil prices, more than 30 percent of electricity generation in OPEC comes from oil-fired generating plants. By contrast, virtually no electricity is produced in North America from burning diesel fuel anymore. That's because if a North American utility wanted to start producing electricity from burning oil, it would have to charge huge amounts for its power to cover the cost of what only recently was as much as $147-per-barrel oil. But when a utility company in Saudi Arabia wants to burn oil to generate power, it's not paying the same fuel price a North American utility would have to pay. Instead, it's getting an even more egregious price subsidy than Saudi motorists get at the pump. Since the spring of 2006, a royal decree has fixed the price of oil for Saudi power plants at forty-six cents per million British thermal units (a standard unit of energy measurement). Since a barrel of oil contains 5.8 million British thermal units, the royal decree translates into a fuel cost of $2.70 per barrel for the Saudi oil-fired utility. That works out to roughly six cents a gallon, a fraction of what even Saudi motorists pay at the pump.

All of a sudden Saudi's massive oil subsidy becomes a massive power subsidy. And just as underpriced gasoline is overconsumed, underpriced electricity is also overconsumed. Power prices in the Middle East are a fraction of those in North America and, not surprisingly, power demand there is growing at a multiple of power-demand growth rates in North America or the rest of the OECD. While Canadians are asked to turn down their air conditioners to conserve electricity, the Mall of the Emirates in Dubai features a five-run indoor ski hill complete with gently falling snow and parka-clad shoppers sipping hot chocolate while the desert sun blazes outside. In countries like Saudi Arabia, where power demand is surging at a 7 percent annual growth rate, blackouts have become commonplace; they have prompted the construction of huge new oil-fired generating stations with the capacity to burn millions of barrels of $2.70-per-barrel oil up their smokestacks.

And as if huge power subsidies weren't enough, there is yet another powerful dynamic driving the Middle East to burn more and more of its own oil. Saudi Arabia isn't just running out of oil; it's also running out of water. So much so that the kingdom has recently rescinded agricultural subsidies, which have diverted an increasingly scarce water supply to farming. Instead, it plans for the country to become increasingly reliant on food imports in order to conserve increasingly precious water resources. But that's not all the kingdom has planned.

As Saudi's ancient water aquifers deplete, the kingdom will have to continually ramp up water production from what are already the world's largest desalination plants. But desalination plants carry huge energy costs. While Saudi will try to burn as much natural gas as possible to power those plants, the country is not nearly as well endowed in that hydrocarbon. Ultimately it will have to burn more oil, in which case water depletion will only hasten the pace of oil depletion in the world's leading producer of crude oil.

What makes the growing domestic thirst for oil in OPEC countries so ominous to Western consumers is that it is coming precisely at a time when oil production in those very countries is either peaking or declining. The tandem of soaring domestic consumption and stagnant production is a recipe for world shortages since it means that major oil-producing countries may soon be cannibalizing much of their own export capacity, at the expense of world oil supply. Even more disconcerting is the prospect that this displacement of exports is likely to occur in the very countries that Western consumers have been repeatedly assured will provide most of the world's future supply growth.

The cannibalization of exports is already apparent. Despite the still-overriding importance that the world attaches to OPEC production, exports for the cartel hardly grew at all over the first half of this decade and will fall over the second half. Had it not been for the simultaneous surge in Russian oil exports, prices would have risen a lot sooner than they did. The cartel has not been able to boost supply for the past five years and is struggling simply to maintain current production levels. Exports will likely fall by nearly a million barrels per day over the next five years and more steeply thereafter as the depletion in many of the world's oldest oil fields accelerates.

The Middle East has been the workhorse of world petroleum production for almost a century and the cumulative draw from the region is beginning to show. Once-huge fields like Burgan in Kuwait are now a shadow of what they were, with production rates at a fraction of their former peak. And there is growing speculation that Ghawar, the world's largest field and home to more than 50 percent of Saudi Arabia's production, may soon start to decline as well. At the same time, the production rate in other Middle Eastern oil heavyweights like Iran is a fraction of what it was in the glory days. Iran's current production of just less than four million barrels per day is almost 40 percent lower than it was during the time of the Shah, and its exports to the world have fallen by well over a million barrels per day from the levels the world enjoyed before the Iranian Revolution in 1979.

The story isn't much different in the rest of OPEC outside of the Middle East. Production in Venezuela, one of the organization's founding members, has been falling steadily, a decline that began even before Hugo Chávez effectively nationalized the country's oil industry. Conventional oil production in the country has fallen by over a million barrels per day over the last decade. Ditto for Indonesia, where production is now just over half of the level it was only twenty years ago.

Indonesia is a particularly intriguing example because it represents Western consumers' worst fears. Indonesia has been an oil-importing country since 2004 and, in early 2008, finally handed in its OPEC membership card. Its million-barrels-per-day of production is now almost 20 percent lower than its daily oil consumption, hence it must now draw upon supplies from the rest of the world, when a decade ago it was exporting oil to those very same places. We used to burn their oil; now they burn what might have been ours.

Depletion is challenging enough in its own right, but when you marry it with runaway domestic fuel-demand growth in oil-producing countries, the problem for world oil markets becomes much more acute. Exports from OPEC have simply not grown over the last five years, and will actually decline by about a million barrels per day over the next five years. Just exactly who is going to compensate for that loss of supply is far from clear.

Up until very recently, the world was betting that if OPEC dropped the ball, Russia would be there to pick it up, just as it has done for the better part of the last decade. Russia is not only the world's second-largest producer of oil but, unlike Saudi Arabia, it is a country where domestic oil consumption had been falling during the economic chaos that befell that country. Not only did Russian oil production soar from 6.7

million barrels per day to almost 10 million barrels per day from 2000 to 2006, but virtually all of the production increase went to exports. Since 2000, Russian oil exports have risen almost 3 million barrels per day, filling the breach created by OPEC's growing cannibalization of its own production. In the process, Russian exports accounted for 70 percent of the increase in supply from the world's key exporting countries between 2000 to 2006. While the world has waited with baited breath for every OPEC supply announcement, the growing exports from Russia, not from Saudi Arabia, have allowed motorists around the world to fill up their tanks. But now Russia's exports are succumbing to the same forces that have curtailed OPEC's exports.

Russia's energy officials have already admitted that Russian oil production did not grow in 2008, and would likely not grow at all over the near term due to depletion in the mature oil fields of the Western Siberia basin. At the same time, efforts to expand production in Eastern Siberia have been met with limited success, with the region's largest project, Sakhalin 11, mired in controversy and mammoth cost overruns.

But, as is the case elsewhere, depletion isn't the only factor constraining Russian exports. There has also been a renaissance in Russian energy demand until the recent global recession, as befits a booming resource- and energy-rich economy that has been growing at 6-percent-plus every year. After years of decline, Russian oil consumption rebounded to grow at a very robust 4 percent annual rate, on the back of a spectacular boom in auto sales. Russians are now burning more Russian oil, and that means less for the rest of us.

As in OPEC, oil production is no longer expected to grow in

Russia, and with the recent resurgence in domestic demand, oil exports will fall by more than half a million barrels per day by 2012. Instead of offsetting the decline in exports from OPEC, as Russia has done in the first half of this decade, the northern energy giant will now exacerbate the loss of OPEC exports with curtailments in its own oil shipments to the rest of the world.

Together, the two largest oil producers in the world, OPEC and Russia, will not only be unable to meet any future growth in global oil demand, but, by 2012, between them they will be cutting back nearly one and a half million barrels per day of supply to world oil markets. We will be asking them to put production in high gear to meet rising demand, but they will be throwing it into reverse. And the coming supply crunch doesn't end there. Another oil-exporting country is likely to lose more of its exports than either Russia or OPEC over the next five years.

Exports from Mexico, until recently the number two supplier behind Canada to the huge U.S. oil market, will virtually collapse over that time frame. While Mexico is not going to become a net oil importer like Indonesia, it will nevertheless cease to be an oil exporter of any consequence. Its current 1.5 million barrels per day of oil exports, virtually all to the American market, will shrink to something like a quarter-million barrels per day or less.

Mexico's problem is that its major oil asset, the vast Cantarell field under the Bay of Campeche in the Gulf of Mexico, is just about tapped out. Currently the third-largest producing field in the world, it is home to just under half of Mexico's oil production. But like other underwater fields such as those in the North Sea, Cantarell is drying up quickly.

It has already lost over a million barrels per day in production and it is expected to lose at least another half-million barrels per day over the next five years. Juxtapose that supply outlook against steady growth in domestic gasoline consumption and, within a half decade, pretty well all of the country's depleted oil production will go to filling the tanks of Mexico's growing fleet of cars.

And that is not even the end of the bad news. Oil optimists often make the case that it is not a geological limit that is choking off supply, but static or declining capital investment in oil production. There is plenty of oil in the ground, they say; we just aren't getting it out fast enough. But the other side of this coin is the argument that even if there is sufficient oil in the ground to keep the world economy running, we wouldn't be able to exploit it. The fact is that there are simply not enough oil tankers, drill rigs or refineries out there to meet demand. The investment required to keep the oil flowing runs to the trillions of dollars, a level and a rate of investment far greater than what we are seeing. Whether the problem is that the state-run oil companies in the producing nations are corrupt, or inefficient, or technologically backward— or whether they know that there is little point in investing in infrastructure when the oil is running out—is beside the point. And with the recent decline in oil prices brought about by the global recession, it is these state-owned behemoths that are leading the cutbacks in investment. The best opportunities for expanded production appear to be in places where it is unlikely to happen.

Together OPEC, Russia and Mexico account for more than 60 percent of world oil production, with combined output of just over forty-seven million barrels per day and total exports of roughly thirty-five million barrels per day. While the group as a whole should be able to maintain total production rates close to current levels, growth in domestic consumption will cut exports by more than two and a half million barrels per day. That's more than enough of a squeeze on future oil supply to see oil prices back again in triple-digit territory over the next business cycle.

Implications for Canada

While the world's major oil-producing nations are approaching peak or looking back at it wistfully, hopes for new supply from unconventional sources are pushing new players to the fore. Canada in particular is one of the very few places that can not only expand production but can expand exports to the U.S. as well. While Canada's conventional oil supply, found mainly in Alberta's Western Sedimentary Basin, is depleting as rapidly as any other mature basin in the world, its vast oil sands deposits give it a production upside that few other countries have.

As less and less oil will flow from the places that we have depended on in the past for fuel, more and more will have to come from these new frontiers of world supply. That is already an extremely costly and problematic frontier, and scarcity will make it even more so in the future. Today oil sands are the boundary of that frontier, tomorrow it may be oil shales.

The one common denominator of the new sources of oil supply is that, unlike the conventional fields they are replacing, they all require a tremendous amount of energy to get the oil out of the ground. And each new source of supply seems to bring a higher and higher price tag. It cost as little as $10 per barrel to bring new supply out of the North Sea when OPEC shut off the taps over three decades ago; it was this abundance of cheap oil that got the global economy back on its feet after the oil shocks of the 1970s. But today new supply from a source like the oil sands will cost as much as $90 per barrel (so that is where global prices will have to be before a new oil sand project gets started).

That's precisely why it will take dwindling exports from traditional suppliers like OPEC and Russia to bring new oil sand production to the market. And that's exactly what is going to happen. The imminent collapse of Mexico's oil exports to the U.S. marketplace together with the growing impairment of both the Middle East's and Russia's export capacity all but assures us that we will see those oil prices and even higher ones again. When we do, will see a new round of development of Canada's 165-billion-barrel resource. Scarcity will bring high prices, and high prices will make Albertan synthetic crude economically viable.

But as we substitute synthetic oil from Canadian or even Venezuelan oil sands for dwindling supplies of Sweet Arab light crude or other low-cost conventional oils, we not only climb up the cost curve, we also climb up the carbon curve. You have to burn an awful lot of natural gas to get oil from oil sands.

As more of the oil from places like OPEC is burnt up as 25-cent-per-gallon gasoline pump prices or 7-cent-per-gallon power plant fuel, the rest of the world has to scrape the bottom of the world supply barrel. And if that stuff looks like guck coming out of the ground, just look at the carbon trail that extracting it and processing it leaves up in the air. The global demand shift in energy is leading us to ever more carbon-intensive sources of new supply. Producing a single barrel of synthetic coil emits three times as much carbon as producing a conventional barrel of oil.

And so Canada will find it looks more and more like the solution to one problem, but that it makes another problem worse. Canada will have a unique role in the coming decades as both hero and villain, a role thrust on it as other oil-exporting countries leave the stage.

William Marsden
THE PERFECT MOMENT

Author and documentary-maker William Marsden is one of
Canada's foremost investigative journalists. His most recent book,
the national bestseller *Stupid to the Last Drop: How Alberta Is
Bringing Environmental Armageddon to Canada (and Doesn't Seem to
Care)*, won the 2007 National Business Book Award. He is the
winner of two National Newspaper Awards for investigative
journalism and three Prix Judith-Jasmins for journalism in
Quebec. He is also co-winner with the International Consortium
of Investigative Journalists in 2003 of the U.S. Investigative
Reporters and Editors Award for an investigation into the
privatization of water published by The Center for Public Integrity
in Washington, DC.

It is always a little hard to find a convincing answer to the man who says, 'What has posterity ever done for me?' and the conservationist has always had to fall back on rather vague ethical principles postulating identity of the individual with some human community or society which extends not only back into the past but forward into the future. Unless the individual identifies with some community of this kind, conservation is obviously 'irrational.' Why should we not maximize the welfare of this generation at the cost of posterity?

—Kenneth Boulding,
"The Economics of the Coming Spaceship Earth," March 8, 1966

Soon after he became Alberta's premier, Ed Stelmach quickly set everyone straight. There would be "no touching the brake" on the tar sands, he announced. For Stelmach the tar sands are a hulking luxury vehicle with a powerful, throbbing engine, and the way to drive it is simple. You crank the ignition, point the beast in any direction you want and let her run. The only rule is that once you hit the road there will be no easing up on the gas. As the Lexus ad says: "This is it. This is what a perfect moment feels like. Brilliant. I love it." This is Alberta's moment. Fuck posterity.

It's no mystery why Canadians would rejoice at the mere thought of the tar sands, mesmerizd as they are by such a colossus. The wealth encased in those tiny black grains is staggering. The expanse of the sands is enormous. They underlie an area, as the Alberta government website boasts, "larger than the state of Florida, twice the size of New Brunswick, more than four and a half times the size of Vancouver Island, and 26 times larger than Prince Edward Island." We have barely dug up 2 percent and you can already see the scars from the moon, or so I'm told. As conventional wells dry up and oil companies spend increasing amounts of money and energy to find less and less oil, the certainty that there are 174 billion barrels—and possibly a lot more—in northern Alberta should be heartening.[1] That's enough to meet Canadian needs at present consumption for the next two hundred years at least. As prices hit a record US$146 a barrel in July 2008, tar sand reserves added up to more than US$25.4 trillion. It's made our gross domestic product figures sparkle. Our public debt is the lowest among the G7 countries. Our employment figures, in Alberta at least, have soared as jobs go begging and we're forced to bring in thousands of foreign workers. Literally every gauge of wealth and good fortune seems to show that we've got nothing to worry about, thanks to the tar sands. The world financial crisis that struck in October 2008 may have slashed energy stocks in half and reduced the high price of oil to less dizzying heights, but it did not affect the fundamental profitability of energy companies' operations, which continued to gush cash. We can carry on down the endless highway of our dreams, firmly convinced that the world as we know it will go on forever.

Don't you just wish it were so?

The world as we know it is essentially a blip on a geological timeline—the brief but spectacular "Age of Oil." The unrivalled flexibility and stored power of fossil fuels has allowed us to escape from a world that held us hostage to the limits of our muscle fibre. The energy stored in one barrel of oil is equivalent to five labourers working 12-hour days non-stop for a year. The offering, however, is much more than brute force. Our machines not only do the grunt work for us but also are "manifestations of a perfect life, assisting you to realize goals, achieve greatness, and passionately embrace every perfect moment." That's what the Lexus ls460 brochure tells us. (But we don't believe that, do we?) Fossil fuels have shaped every facet of our lives, right down to our thoughts, hopes and dreams. These hopes and dreams are moulded with petroleum oil, just as plastic is.

One of the obvious ironies our world ignores is that we refuse to extend our pride of mathematical precision to the basic math of world oil reserves. It's simple. Our reserves total about 1.2 trillion barrels, give or take a few hundred million. We use this oil up at a rate of more than thirty billion barrels each year. That means world reserves will last about another thirty-nine years. Petroleum geologists, such as Andrew Miall at the University of Toronto, predict that what's left undiscovered will give us another ten to twenty years, tops. When I pointed out this dilemma to a friend, an intelligent, fairly well-informed self-made millionaire, he quite innocently asked: Can't we just make more?[2]

The harsh reality we are waking up to will reek like the stench of the tar sands. The fact that we are chewing up a billion years of energy accumulation in a mere two centuries, and that our world is not capable of infinite growth, will inevitably shatter our delusions and force us to accept that the world is a finite, closed system of limited resources that has to be managed not by greed but by thought. The fact that we are so heavily mining about one-fifth of the entire province of Alberta to extract a dirty, climate-changing fossil fuel is not a reflection of victory. It is a sign of defeat. Drilling and pumping is easy and cheap compared to the expensive and often difficult process of extracting oil from the tar sands. Yet every major company is spending billions for the privilege, their faith in the drill bit shaken. Like a frightened army, pillaging and looting in its desperate retreat, we are destroying a vast ecosystem vital to our long-term well-being just to keep the perfect moment alive.

The tar sands suddenly are a root metaphor for every pressing issue we face both as Canadians and as members of the human species. We are caught in a confused state of hope, denial and necessity: hope that our technological brilliance will allow us to uncover a clean energy equivalent to oil just in time to keep the perfect moment going; denial that the more likely outcome is that we face drastic changes in the way we live; and the bleeding necessity to keep burning oil and gas because it propels us to our work stations and it keeps us from starving or freezing in the dark. Someone once said we are four missed meals away from chaos. That was then. Now it's one tank of gas.

I first visited the tar sands late in the fall of 2006. At the time, Syncrude Canada Ltd., the largest operator in the sands, was expanding its operations to increase capacity to 350,000 barrels a day (15 percent of Canada's daily consumption), from 300,000—enough to keep the world chugging along for six minutes. Given the complex systems that have to be put in place, fine-tuned and coordinated, that's a huge undertaking. In places like Saudi Arabia, oil companies just pump crude out of the ground at a cost of about two dollars a barrel and transport it to refineries around the world. Pump jacks do most of the work. It's hardly even fair. In northern Alberta, you have to dig up the sand, mix it with hot water, pipe it to an extractor that separates the bitumen from the sand and then pipe it to the upgrader/refinery where it is turned into crude. What's more, you have to do it all in a climate that can be brutally cold and unforgiving. The bottom-line cost varies depending on how good you are at keeping the machine turning. For Syncrude the costs jump between fifteen and twenty-five dollars per barrel. But with oil prices bouncing around on an overall upward joyride, there's nothing to worry about. There's a hefty cushion.

Profits, however, are not the point. The issue faced by consumers is the flow. The volume of oil reserves in the tar sands is approximately the same as the non-OPEC reserves. Non-OPEC reserves have been flowing at nearly 50 million barrels a day. Flow from the tar sands, after thirty years of heavy investment, is 1.4 million to 1.7 million. And that's the problem. There is a huge volume of oil in the tar sands. But we have so far found it impossible to flow that at any significant rate without totally destroying the environment, although we are doing a pretty good job on that line. The tar sands mirror all those dirty hydrocarbons such as shale and coal that carry too high an environmental price to exploit even as we run out of conventional crude. So when oil men say there are plenty of hydrocarbons in the earth, they're right. But it's no more useful than you sitting there starving and being told that farmers are going to grow a lot of corn next year. It's just not relevant.

My guide, Peter Duggan, and I trundle along dirt roads in his pickup truck through the cool autumn air past towering refineries with their shiny tangle of pipes and chimneys belching steam and flaring off gas, then along the edge of two open-pit mines that measure about thirty kilometres from end to end—Syncrude's eight tar sands leases total about one hundred thousand hectares, almost twice the size of Toronto—before descending into a third mine that had only recently been opened. It is approximately seven kilometres long with an adjacent toxic tailings pond measuring three kilometres by three kilometres, which in my book makes it a toxic tailings lake. It's in these massive lagoons that the oil companies dump the poisonous, chemical-laden detritus of their vast processes. About eighteen barrels of the stuff for every barrel of oil produced. This is where five hundred ducks died in April 2008 when they landed on the oil-soaked ponds. Syncrude claimed it was an isolated incident. But I was told that migratory birds die in those ponds all the time. Local Cree are employed to rake them out. The entire Athabasca Delta and surrounding boreal forest are nurseries for hundreds of species of song and water birds. Forest destruction destroys their breeding grounds. The tar sands kill them.

The ponds, some of which are held together by the world's second-largest system of dams, are a mixture of water, sand, silt and bitumen, a tar-like mixture containing hydrocarbon toxins. They are breeding grounds for methanogenic bacteria—bacteria that produce methane, a greenhouse gas that is at least twenty times more damaging to the atmosphere than carbon dioxide.[3] The oil companies are studying ways to reduce the lighter hydrocarbons in the tailings, thereby reducing the munching bacteria and their "fugitive gases."[4] But, in the end, nobody knows what to do with the tailings. The companies hope that bacterial action will eventually dispose of the hydrocarbons. Some scientists believe that freezing them into ice pellets will reduce their toxicity. But there's no sign of a speedy neutralization. The only certainty is that the ponds will continue to expand, swallowing up more and more of the boreal forest.

Each barrel of oil extracted produces on average three cubic metres of tailings.[5] At today's production rates, that means about 1.23 billion cubic metres of tailings are blasted annually out of fat steel pipes and into the ponds in a furious shower of discarded oil and sand. By 2020, the ponds are projected to grow to 220 square kilometres from the present fifty. That would make them equivalent to Alberta's larger lakes. The tailings water has been leaching into the watershed at an estimated rate of 11,000 cubic metres a day. And there have been incidents of serious spills flowing downriver to the Athabasca Delta, where the Cree have found that their fish occasionally smell of burnt rubber or have strange lesions or mutant growths, such as two mouths.

Studies done on behalf of the oil companies have found unsafe levels of arsenic, mercury and polycyclic aromatic hydrocarbons in Athabasca Delta sediment, water, marine plants, fish, moose and other mammals. A study done by biologist Kevin Timoney in 2007 for the local health authority in Fort Chipewyan, a community of about 1,200 Cree, Chipewyan Dene and Metis on Lake Athabasca near the delta, concluded that "concentrations of these contaminants, already high, appear to be rising."[6] According to local doctors, cancer rates in Fort Chipewyan have soared in recent years. Both federal and provincial health officials have denied there is anything unusual about the cancer rates. It's a denial based on nothing. They then tried to get a doctor, who had rung the alarm, fired by complaining to the Alberta College of Physicians. The initial charge was that he had unduly raised fears among the Fort

Chipewyan population. Then they accused him of double billing. The College dismissed the charges, but by then the doctor had decided he had had enough of Alberta and moved to Nova Scotia. Neither federal nor Alberta health officials have done a comprehensive study of the diseases that are afflicting the people of Fort Chip. In June 2008, they promised a new study but have not followed through.

Peter and I climb out of his truck to take in the breadth of what miners call the "footprint," which is the physical scar we leave on the crust of the Earth after we have bulldozed, driven out or killed off the so-called overburden, which in this case is the boreal forest. It includes topsoil, wetlands, fens, streams, peat bogs, trees and of course wildlife. This boreal forest naturally cleans carbon out of the air, purifies water and produces oxygen and nutrients while supporting wildlife that has as much right to be on this Earth as we do. It is a delicately balanced natural industrial plant that helps make life possible on Earth. Destruction of the boreal ecosystem has not only trashed this energy system but also transformed what was once a carbon sink storing billions of tons of carbon dioxide into a carbon emitter.

That's what happens when you kill forests. They surrender their stored carbon.[7] For this reason, the tar sands represent a triple whammy to the environment. First, a forest no longer stores carbon. Second, it emits it. Third, millions of tonnes of greenhouse gases are created by the process of extracting bitumen from the sands and refining it into crude. This process is three to five times more greenhouse gas intensive than conventional oil refining. Against any measure, the figures are staggering. Since 1990, Canadian greenhouse gas emissions associated with extracting, processing and transporting fossil fuel energy to the U.S. alone increased more than 151 percent.[8] Just two tar sands companies—Suncor and Syncrude—account for more than 3 percent of Canada's total greenhouse gas emissions. And they are just beginning. Suncor's expansion plans will more than double its emissions by 2012.[9]

Black hills extend for as far as the eye can see. The heavy smell of oil clogs my nostrils. The only sign of life—aside from Peter and myself, of course—is the machines. Distance is hard to judge in this monstrous landscape. A three-storey-high electric shovel, which costs thirty million dollars, is dumping buckets of tar sands into a ten-million-dollar truck. The machines seem only about one hundred metres away, but really the distance is five times that. The endless moonscape reduces them to Tonka toys. Each lumbering truck carries a payload of about four hundred tonnes, which miners call "the feed." That's where the money is.

The money should explain all of this—the bulldozed forests, the lakes of poison bobbing with the corpses of ducks and other wildlife while silently releasing their invisible gases. In terms of conventional economics, it constitutes a very big payoff. One truckload will be boiled and cracked into two hundred barrels of Syncrude Sweet Blend (as good as the industry standard, West Texas Intermediate). At the 2008 record price of us$145 per barrel, that one truckload is worth $29,000. At Syncrude's planned peak capacity of 350,000 barrels per day, that would make 1,750 truckloads carrying $50.7 million of crude per day or about $18.5 billion per year. Roughly speaking, the company now produces crude at a rate of about three barrels per second. As of October 8, 2008, at exactly 1:11:53 p.m., Syncrude had produced 1,893,000,290 barrels of oil since the company started production in July 1978.[10] At 2008's record price, that is $274 billion. But as we have come to learn, in today's skittish world, the price of oil sinks and rises like those dead ducks. Right now it has sunk to $48.86, which means Syncrude is earning $9,772 per truckload. Not to worry. They are still very much in the game.

People will do just about anything for a few hundred billion dollars. But even at that price, is it all worth it? Almost all of that oil has been piped down to the United States through one of Canada's three western trunk pipelines. Since Syncrude started digging bitumen out of the ground, American consumption has increased 25 percent.[11] They have 4.5 percent of the world's population but consume 25 percent of its petroleum. All major Canadian pipelines lead to the U.S. We are now their largest single source of fossil fuel energy, supplying about a quarter of their petroleum imports. At the moment we have no choice. The North American Free Trade Agreement (NAFTA) basically allows Americans to take what they want. We could cut back, but we would have to maintain the percentage—about 60 percent—we are now exporting south. In a world running out of oil, power has shifted from the consumer to the producer. But not in Canada. We surrendered our producer rights when we signed NAFTA and the only way to get them back would be to get out of NAFTA.

We live in a cold country that is running out of conventional oil and gas. Alberta's conventional oil production peaked in the 1970s and has declined 40 percent since about

1995. Our natural gas reserves have been declining since the mid-1990s. Millions of Canadian homes and businesses rely on natural gas for heating and as raw material for industrial processes such as manufacturing the fertilizer used to grow our food. Yet we continue to ship most of what we produce directly into the U.S. From 1990 to 2006, oil and gas exports to the U.S. increased 154 percent, which is about twice the rate of growth of natural gas production and three times that of oil production. Most of the increase in natural gas exports occurred between 1990 and 2000. The only reason we haven't exported more gas is because we simply no longer have the capacity. We are consuming and exporting faster than we are replenishing our reserves. In 1985, we had enough gas reserves to last thirty-five years. As of 2008, we have enough to last only nine years. "Demand for natural gas is expected to rise, driven by the oil sands industry," the U.S. Energy Information Administration 2008 report on Canada states. At this rate, Canada could have trouble meeting its own needs by 2015, or even sooner.

But this hasn't stopped oil sands companies from laying plans to double their natural gas consumption. Among the many rancid ironies that accompany tar sands exploitation is the fact that we burn the cleanest fossil fuel—natural gas—to produce bitumen, which is the dirtiest. The oil sands use enough natural gas each year to heat more than one million Canadian homes. Oil sands companies boast that they have decreased their greenhouse gas emissions. But most of this decrease is owed to an increase in the use of natural gas.

The next twelve years will see production almost quad-ruple. Tar sands production is expected to increase to as much as 4.3 million barrels per day by 2020, from the current 1.2 million. So we will be producing every year almost as much as Syncrude has produced in the last thirty years. "This degree of activity would . . . see Alberta become a Global Energy Leader," Alberta Energy's website boasts.[12]

Both Canada and the U.S. are gearing up for the increased oil sands production. Over the next ten years, we will have doubled our pipeline capacity into the U.S. market. In preparation, U.S. refineries are expanding and retooling to handle 800,000 barrels per day of bitumen (66 percent of our current production) from the tar sands for processing into crude oil and other petroleum products that add value to the resource.[13] We could have a scenario where America is selling refined petroleum products made from Canadian tar sands back to Canadians. It's like exporting logs and importing fine furniture. But then, we do that too.

It would be some comfort to believe that we are making so much money out of all of this that we are building a war chest that will help us through what will inevitably be difficult times as climate change worsens and the world begins transforming itself from an oil-based economy to one based on, well, nobody really knows. But so far there's no basis for that comfort. We are making money. But we are spending it just as fast.

Since 1971, well over two hundred billion in constant dollars has flowed into the Alberta treasury from resource development. All of it has been spent. Alberta's Heritage Fund, which was established in 1976 to park money for future generations, has succeeded in accumulating only $16.6 billion. That's not very much when you consider that the fund started with a $1.5 billion contribution and a promise from the province to commit 30 percent of the non-renewable resource revenues to the fund every year. That promise was never kept. One reason is that the province endured tough times during the late 1980s and early 1990s, when the price of oil and gas plummeted. But that's a flimsy excuse. The real reason is the Alberta Conservative government has consistently cut taxes to buy votes, using the oil money to pay its day-to-day expenses. Even during the boom times, which have lasted now for more than ten years, the province has spent practically every cent it has earned. Consider this: if the province had obtained an average 10 percent return just on the first $1.5 billion, without putting a cent more into the fund, it would now be worth $39 billion. It would be earning almost $4 billion per year in revenue. Let's go further and pretend that each year the government parked $1 billion into the fund from its oil and gas revenues, which would be far less than the promised 30 percent annual contribution (and considerably less than its resource revenues for 2007–8, which were $11 billion). At an overall return of 10 percent, the fund would now total $246 billion and would be gushing about $24 billion a year in income revenue—equivalent to almost two-thirds of Alberta's annual expenditures. In other words, it would be more than enough to finance Alberta's entire provincial health and education budgets and still have ample cash left over for the paltry $403 million it spends on its environment department, which oversees the environment's destruction. Far from building the Heritage Fund, the province continues to loot it. Despite the billions of dollars rolling into its treasury owing to record high oil prices, in 2007 the government transferred $180 million from the fund to meet its general revenue needs. Consequently, there was a net decrease of $13 million in the fund's value, according to the fund's 2007 annual report.[14]

Compare Alberta's oil fund with Norway's and you get a clear picture of how an oil economy should be managed not just for the well-being of the present but also for future generations. Even though Norway's fund has been operating only since 1995, by the beginning of 2008 it contained $412 billion (or about $80,000 per capita compared with Alberta's $5,000). The reason is simple. Norwegians took control of their energy resources in the North Sea at an early stage. Realizing the obvious implications of an economy based on a non-renewable resource, the Norwegians wanted to build for the future by making sure that the profits stayed home. Norway's Statoil company has at least a 50 percent interest in every energy project. On top of that, Norway has high royalties. To counter the inflationary pressures that inevitably accompany a high-flying oil economy, the fund invests only in foreign assets. Not a penny is spent at home. Norwegians understand that if they start throwing money around in their own backyard, inflation will quickly gobble up the profits.

Alberta has done exactly the opposite. The Tory government sold off the Alberta Energy Corporation, which has since morphed into Encana Corporation, one of the largest and most profitable oil and gas producers in Canada. It imposed the lowest royalties in the world, thinking that that would encourage investment. And it lowered taxes so that the electorate would keep re-electing the Conservatives. This has worked out well for the Conservative government. But not for Alberta. Or for Canada.[15]

The 2008 economic survey of Canada produced by the

Organisation for Economic Co-operation and Development (OECD) cautioned that Alberta's boom is risking the health of the national economy and costing jobs in the manufacturing sector. Inflation has become so high that, as the *Globe and Mail* reported in June 2008, the trickle-down wealth is not happening as planned. "An overheated economy has driven costs in Alberta sky high, making it less enticing for companies to ramp up investment in the province," the paper's Report on Business section stated. Without the reinvestment, which in Alberta was twenty billion dollars in 2007 alone, what real value is anybody getting out of the oil sands? The Alberta government spends almost every penny it earns to meet its annual budget expenditures. The trade-off for Albertans is that they have the lowest taxes in Canada. To the average Albertan, living a boomtown moment is enticing. It's a chance to make a stake. Work the sands ten days straight, twelve-hour shifts, and then fly south to the family for a week of R and R, clearing fifteen thousand dollars a month. Save your money and within ten years you could retire. Who would ever vote against that? But it rarely works out as planned. Such is human nature. The stake is never big enough. The costs keep rising and the money rarely stays in the bank. Too many toys and necessities to buy. Particularly for newcomers. The inflationary spiral that has a trailer home selling for five hundred thousand dollars in Fort McMurray is simply eating up their high salaries. By the end of 2007, Alberta's inflation rate was three times that of Ontario's. But the rate itself is deceptive. It does not take into account construction costs, or the costs of building a school or a road. The cost of this infrastructure has risen far higher than the so-called basket of household goodies upon which we base our inflation rate.

To stop the inflationary and negative currency exchange effects of such unrestrained development, the OECD survey had this advice: "Alberta should implement allocation and withdrawal rules for its Heritage Fund: preferably, it should save all its oil revenues in a foreign asset fund, as Norway does, spending only smoothed yearly fund income. The federal government should consider doing likewise." Norway's Statoil company has bought leases in the tar sands. Canadians could see the day when Norwegians are making more money out of our sands than we are, and exporting inflation to us in return.

Alberta has essentially abandoned the field to private companies and their investors, most of whom are foreign. The result is that most of the profits from our most precious nonrenewable resource are leaving the country.

While Alberta raised its royalties in 2007 about 20 percent overall—a piddling increase when you consider the tiny amount oil companies had been paying—the government sadly reversed itself when the financial crisis hit in October. Albertans no doubt will soon resurrect the 1980s bumper sticker begging God for another oil boom and promising not to "piss it away this time." For each of the four fiscal years after 2003, oil sands operators returned a little over one billion dollars to Albertans. It's difficult to calculate what these companies' gross oil sands revenues were during that period, because they are not all reported. But in the case of Suncor, which is the second-largest tar sands operator, the company in 2007 paid $691 million in royalties on gross oil sands revenue of $17.9 billion, or 3.9 percent. Basically, Albertans are giving the oil away for chump change like some endless Boxing Day sell-off. What's more, Alberta awards oil sands companies an income tax deduction against royalties. Suncor in 2007 paid a mere $513 million in federal and provincial income taxes. How many Canadians get to pay only 2.8 percent taxes on their gross income? And tar sands developers are among the most profitable companies in Canada. Suncor's five-year return has averaged 30 percent while Syncrude's main shareholder has earned 50 percent.

Canadian companies may be leaders in the oil sands, but a large portion of the profits go to foreign investors. Syncrude, which owns the largest and richest leases in the Athabasca oil sands region, is 45 percent owned by foreign companies, most of which are American. Many of the investors in Syncrude's Canadian-owned companies, such as Canadian Oil Sands Ltd., which holds 36.74 percent, are foreign. Canadian Oil Sands is a trust, which means that money paid to foreign investors is tax-free. Of the sixty-four projects in the Athabasca tar sands region, foreign companies control nearly half. The list of countries signing up to spend billions on the tar sands is an indication of how important they have become as one of the world's key energy banks: Norway, Korea, the U.K., Netherlands, France, Italy, China, the United States and Japan. We are incurring the costs, and the benefits are migrating all over the planet.

So, what am I thinking as I stand in the pit trying to absorb my surroundings? I'm surprised at my own answer. I'm thinking that this is the way life should be: filled with great human endeavours unencumbered by self-restraint, focused on the universal task of extracting energy. That is what mankind has always done and always will do, restrained only by the nature of the energy source. Yet the thought that this level of destruction represents normality is horrifying. But I soon realize that I feel this way because when you are in the pit, the pit is all you see and you can't imagine life any other way. You have to rise up ten thousand metres or more and circle the widening sacrifice zone of your psyche to get, so to speak, the whole picture, and see the wild beauty of what once was and will never be again. Only then can you even begin to understand what is happening.

Conventional economics puts no value on what is lost. Economists and accountants refer to environmental devastation as a "negative externality." Extracting bitumen, refining it into crude oil and selling it is simply business as usual. Those who take part in that deal are the beneficial parties. The resultant environmental destruction is not part of the deal and therefore is "external" and somebody else's problem. Somebody else gets to absorb the cost, pay the piper. A bit like civilian casualties in war. Presumably the economists, accountants and businesspeople who make the decisions that are changing the planet have no innate lust for destroying landscapes and habitats, unless of course they are insane. But the impact of the destruction of the "overburden" has no effect on their calculation of profit and loss, other than the cost of running the bulldozers that sweep away the overburden, which is chalked up as an expense. In other words, the existence of the boreal forest, with its streams and flora and fauna, registers only as a negative outside the deal-making process of selling oil. The reason is simple. An ecosystem is not a marketable item. You can break it down into its parts—logs, peat, topsoil, bitumen— and sell that off. But the system itself has no inherent value.

You can't sell a natural system, at least not in our society. So it is by definition worthless.[16]

But let's say that we designed a system of accounting that reckoned the value of those things that cannot be sold. We often refer to Nature as something so valuable that you cannot put a price on it. We'll say something like: "Getting out in the fresh air, that's priceless." Or, "You can't put a value on good health." The problem is that by not putting a price on these externals, as our history has repeatedly demonstrated, we open the door for their destruction because eventually everything is subservient to the marketplace, including human well-being.

So let's say we were to assign a value to the benefits we enjoy thanks to the boreal forests, such as clean water and clean air. One thing is for certain: oil from the sands would be a lot more expensive than it is today. It would be priced out of the market. But why stop there? What about the cost to the well-being of societies that for thousands of years have existed in the area? There is no question that the oil sands development has destroyed the cultural traditions of native communities such as Fort MacKay, which is now surrounded by tar-sands mining. What does a way of life cost? We should probably find out, since that is what we are burning when we fill up our tanks.

Pricing these externals is not difficult. There are a growing number of ecological economists in Europe, the United States and Canada who have created viable models for doing just that. One of them is Robert Costanza of the Gund Institute for Ecological Economics at the University of Vermont. "What are the external costs of developing the oil sands and what is the net energy once you have taken all those external costs into account?" Robert Costanza asks. "Damage to the water required for processing the oil sands [and damage to] the land area used for production because it's not that those lands were not producing valuable services [watershed protection and carbon sequestration, to name only a couple] before the oil sands production. So the analysis one would do is a sequence of comprehensives that would show if the oil sands were a net yield at all. It would help [us] to understand what the real net value is."

If you believe in what American economist and social scientist Kenneth Boulding called a "cowboy" economy, then you don't generally care what kind of mark you leave. You take what you can and you ride on. There are endless new horizons. What's left behind is left behind. But if global warming has taught us anything, it's that life on Earth is more complex than simply putting the pedal to the floor. It's taught us what we have known all along but have chosen to ignore because we're having too much fun. It's taught us that everything has its limits and that we have reached those limits and exceeded Earth's carrying capacity. Our footprint has become too big. The reality is that we live in what Boulding calls a "spaceship" economy, with closed systems that have to be carefully maintained in order for life to be sustained. As humans, we have simply ignored—and to a degree not really understood—the crippling effects our actions have wrought on Earth since the Age of Oil put so much energy at our disposal. We have been thinking of our assets as liabilities, and our liabilities as assets. To change our economics would mean a total change in our lifestyle, and there's no sign that we're prepared to do that. But if we don't change our lifestyle, climate change and fossil fuel scarcity will do it for us.

Not surprisingly, we Canadians increasingly live in a confusion of values. A 2008 survey by the *Globe and Mail* found that while 79 percent of respondents said the tar sands are good for Alberta and Canada, more than half of those respondents (55 percent) said that the sands were not good for the environment. The obvious contradiction can be justified only by minimizing or disconnecting oneself from the importance of Earth's ecosystems. The problem is that global warming and the rapid dying out of species makes this level of self-deception increasingly dangerous. We are suddenly caught in the headlights of a new fear: when will it be our turn?

Rick George, president and chief executive officer of Suncor, was clearly worried about the *Globe* poll when he responded with an article designed to reassure the public and his shareholders alike. "I have always been a strong advocate of environmental stewardship and a vision of sustainability [*sic*] that stressed the importance of combining strong financial results with equally strong social and environmental performance," he wrote in the *Globe and Mail* in June 2008. But is that possible? Is it possible for a company like Suncor to operate and at the same time be good stewards of the environment? Of course not. George can talk about how his company has reduced water use and sulphur emissions and state that "Going green makes good business sense." But by 2012, his company plans to almost double its output to 550,000 barrels per day, which will more than double its annual greenhouse

gas emissions to 25.5 million tonnes of carbon dioxide equivalent—equivalent to the emissions of about one-fifth of all the
cars in Canada, from 11.5 million. That alone would make up
about 4 percent of Canada's current greenhouse gas emissions.
No matter how you cut it, Rick George is still in the pit. He is
a miner responsible for a company that for forty years has
wreaked and will continue to wreak, for at least another forty
years, an environmental holocaust in northern Alberta. His
company is a huge contributor to global warming.

Some might say that to blame the Rick Georges of our
world for this destruction would be wrong. He is simply a
smart man who has become a champion in an economic system in which we all partake and benefit. But I'm not sure that's
much of an argument. We may all be part of the system, we
may all be responsible, to paraphrase Rabbi Abraham Heschel,
but not all of us are guilty. I never asked the Rick Georges of
this world to go up to Alberta and destroy a hefty portion of
my own backyard. Nor was I ever asked. Even though global
warming is affecting my life and will drastically affect the lives
of my children and grandchildren, I never had a choice.
Nobody came to me or anybody else that I know of and hon-
estly laid out the pros and cons of tar sands development.
Nobody said that if we build this mine we will destroy this
amount of forest, pollute these waterways, accelerate global
warming and harm Earth's ecosystem in this and that way and
we have no idea whether we can repair any of the damage.
That would have been the beginning of true environmental
stewardship. Waving a green flag after four decades of letting
the free market call the shots does not qualify. If Suncor had
asked my opinion, I would have said no. The price is too high.

If it means less oil, fine. I'll figure something out. I'll campaign to improve public transit and to create compact communities where daily commutes aren't necessary and where all you have to do is walk down to the corner to get milk. We don't have the technology to extract oil safely from the sands, but we certainly have the telecommunications that will permit us to live without relying entirely on the automobile. We'll change our lifestyle and probably be happier for it. We'll find other sources of revenue and jobs. It may not be the perfect moment, but that's okay. I can deal with that.

That's what many of us—perhaps most of us—would have told the Rick Georges of this world, if they had bothered to ask. But they didn't. And that's why we're deep in the pit.

Jeffrey Simpson

BROKEN HEARTS, BROKEN POLICIES:
The Politics of Climate Change

Jeffrey Simpson has been the *Globe and Mail*'s national columnist since 1984 and is a nationally recognized figure and an Officer of the Order of Canada. A Governor General's Literary Award winner, his most recent book (with Mark Jaccard and Nic Rivers) is *Hot Air: Meeting Canada's Climate Change Challenge*.

There are times and issues, as with climate change in recent years, when Canadians have been ahead of their politicians. In poll after poll, Canadians said they wanted more done to combat global warming, but instead they were delivered plans that led nowhere, policies that pursued the paths of least resistance, and a national record second to none for ineffectiveness.

Polls can deceive, however. Responding to a pollster that something should be done does not answer these questions: Action at what price to individuals and the country? Against which other priorities? Over what kind of time frame? And what kind of specific action? Moreover, as skeptics of polling know, the operative point is an issue's salience: Just how strongly does the respondent feel about his or her answer? Does he or she feel strongly enough to frame a vote around the issue? Or is the issue just one among many, and not a terribly important one at that?

Throughout the 1990s, and for perhaps a half decade thereafter, politicians paid lip service to polls saying that Canadians wanted action against global warming. Very few Canadians wrote to their Members of Parliament about it;

few raised it at the proverbial doorstep. The issue was not salient, in other words, except at the margins of the political culture. In the 2004 and 2006 election campaigns, the issue of global warming seldom arose. The parties evidently believed few votes were at stake. The media apparently agreed, perhaps because the media itself was barely seized of the issue.

The Conservative Party's 2006 campaign platform document of forty-six pages, *Stand Up for Canada*, devoted six pages to improving "accountability" in government, four pages to toughening criminal laws, but only one sentence to climate change. The sentence promised that the Conservatives would "address the issue of greenhouse gas emissions, such as carbon dioxide, with a made-in-Canada plan, emphasizing new technologies, developed in concert with the provinces and in coordination with other major industrial countries."

The Conservative platform proposed two specific policies to lower emissions: a federal tax credit for monthly transit passes that would cost the treasury $250 million, and a 5 percent biodiesel and ethanol requirement for gasoline and diesel fuel by 2010. There ended the party's interest in and commitment to addressing climate change.

These promises had much more to do with electoral politics than sound policy. The tax credit would be helpful mostly to those already riding public transit. It would induce very few people to leave their cars behind and therefore have only the most marginal impact on emissions. The corn-based biofuel promise was an agricultural subsidy policy in disguise, since this kind of biofuel is of marginal use, at best, in curbing greenhouse gas emissions.

The Conservatives knew what they opposed: the Kyoto

Protocol, against which their precursor parties (the Reform Party and the Canadian Alliance) had railed. The code words in the Conservative platform—a "made-in-Canada plan"—implied that Liberal governments had allowed inappropriate international norms to be imposed on Canada. "Kyoto is essentially a socialist scheme to suck money out of wealth-producing nations," Stephen Harper had declared when he was leader of the Opposition. In an article for the *Globe and Mail*, he had written, "As an economic policy, Kyoto is a disaster; and as environmental policy, it is a fraud." Harper, echoing the fears of business lobbies, warned against the excessive costs of reducing greenhouse gas emissions, expressed skepticism about the science ("far from settled") behind global warming, and opposed all the Liberal government's measures against climate change. For him, as for the business lobbies, action to reduce greenhouse gas emissions constituted an economic threat, not an opportunity to combat global warming.

After he was elected with a minority government in 2006, Harper appointed to the environment portfolio a neophyte MP, Rona Ambrose from Edmonton. Her lack of political experience illustrated the prime minister's minimal interest in the climate change file. If the Conservatives' platform and their leader's public pronouncements offered any guide, the party was far more concerned about other air pollutants than those that caused global warming. The Conservatives judged that urban smog was a far more salient voting issue than climate change, because smog was something tangible and visible, whereas greenhouse gases were not, and the Conservatives were politically fixated on espousing policies that literally touched the mythical "ordinary, hard-working" Canadian.

The party's heartland lay in Alberta, a province where Premier Ralph Klein had once scoffed that climate change might be caused by people breathing. There were few votes for climate change action in Alberta and, reckoned the Conservatives, not many elsewhere. Canadians seriously exercised about global warming were not Conservative voters anyway. The Conservatives' operating plan for climate change was to go through the motions, nothing more. The Liberals had made much of their commitment to reducing emissions; they emitted five policy papers explaining how they were going to do it, they signed and ratified the Kyoto Protocol, then they fiddled as Canada's emissions rose. Instead of falling by 6 percent from 1990 levels, as mandated by Kyoto, Canada's emissions soared by about 25 percent by 2005. Canada under the Liberals led the advanced industrial world in plans—and in higher per capita emissions. Canadians had seen the Liberals go through the motions and escape political punishment. Why should Canadians respond differently to Conservative inaction?

The Conservatives were wrong. Their carefully calculated indifference to climate change lasted for about six months. It was battered by an aroused Canadian public and, at meetings overseas, by an incredulous international community. Ambrose was hammered in the Commons, embarrassed at international meetings, hesitant and uncertain in committee meetings, and clearly in over her head. In fairness, however, she had been handed the party platform, *Stand Up for Canada*, and was told to defend it, which meant that she was not given much ammunition. She followed her brief to her political extinction, when she was dumped from the environment portfolio into the federal-provincial relations portfolio that is run in its entirety by the prime minister.

In considerable disarray, the Conservatives retreated, regrouped and returned with a set of policies that remained far from ideal but that represented positions no Conservative before the 2006 election would have dreamed possible. Any fair-minded analysis must acknowledge that the Conservatives have come a long way from the moment of their election, not to mention their years in opposition.

The Harper government now favours measures that Conservatives would have fulminated against in opposition: a cap-and-trade carbon trading system, intensity-reduction targets for large emitters, and vehicle emissions standards. There are holes in each of these policies. Large emitters can escape real reductions by putting money into a technology fund; vehicle emissions standards are not as tough as those proposed by California. It is arguable, therefore, whether, even if fully implemented, Conservative policies will reduce Canada's emissions by 20 percent from 2006 levels by 2020, as the Conservatives propose. Similarly, it is important to note that whereas the Liberals had pledged reductions from a 1990 baseline, the Conservatives chose a 2006 baseline, thereby ignoring the 25 percent increase in greenhouse gas emissions during those sixteen years. Using a 1990 benchmark, accepted by the Liberals and the Europeans, means that the Conservatives plan only a 3 percent reduction by 2020.

That the Conservatives suggested any target, given their previous fulminations, represented a sea change in their thinking. The sea change occurred partly because Harper more carefully assessed the mounting scientific evidence and, as a policy analyst, changed his mind. It also occurred because after those early months of disarray, the Conservatives recalculated what was required to satisfy an aroused public. The political culture of the country changed about climate change, so the Conservatives scrambled to adapt. So did other organizations, institutions and parties. To put this observation in the political vernacular used by the Conservatives, climate change went from being a non-issue to being a "shield issue." A "shield" issue, in Conservative language, is one where they can be hurt, as opposed to a "sword" issue, which they can use to damage their opponents. Their climate change policy therefore became a shield. Given their previous attitudes, the Conservatives would never be accorded prima facie credibility on climate change, so it could never become for them a "sword" issue, but they needed some kind of policy to "shield" them from attacks by others.

Seldom has Canada's attention focused so rapidly on an issue. There had been abrupt changes in public opinion before, but around time-specific events such as a constitutional crisis or the 9/11 terrorist attacks. Climate change, by definition, is an incremental issue. Nothing dramatic happens from one day to the next when the climate changes. There are no horror scenes, no deaths or dramatic scenes of the kind required to catch the interest of television, the principal means of communications for the majority of

Canadians. And yet climate change suddenly eclipsed Canadians' perennial number one concern, health care, in 2006 and 2007, before losing some of its drive when the world economy softened in mid-2008.

In the first half of the 2000–2010 decade, and before, room had existed in the Canadian public square for politicians (and others) to hawk anti–climate change messages. That space was vociferously occupied by the Reform Party, the Canadian Alliance and the Conservative Party in federal politics; by the Conservative governments of Alberta and Ontario; by the oil and gas industry; by most business associations; by the *National Post* and the other papers controlled editorially by the Asper family; and, finally, by a handful of scientists who made considerable coin debunking climate change.

As the 2000–2010 decade waned, only the Asper-controlled papers, notably the *National Post* and the *Calgary Herald*, plus a dwindling number of debunking scientists remained in climate change denial. Some of the critics had rather subtly changed their approach. They no longer denied that global warming existed, but they threw up all sorts of slow-down arguments against doing anything. Canada was responsible for only 2 percent of emissions, and therefore even a severe reduction in Canada would mean little in the overall scheme of world climate change. India and China had to move before Canada, since they would be the large emitters of tomorrow. Strong Canadian action would threaten the oil sands, the golden cow of Canada's economy. Canada should wait until a) cities were redesigned, b) higher oil prices brought about lower emissions through declining use of fossil fuels and c) technological breakthroughs occurred. No federal or provincial political party, however, refuted the reality of climate change. Parties started to vie to be perceived as more determined to act than their adversaries. The partisan political culture therefore changed from conflict over whether the problem existed to the best means of addressing it. The climate change deniers were largely relegated to the angry, closed world of cyberspace exchanges, since no one in any position of public authority would listen to them anymore, a spurning that increased their fury but could not stop their decline into irrelevance.

The Liberal Party, whose record on emissions under Prime Ministers Jean Chrétien and Paul Martin epitomized bureaucratic infighting, political sloganeering and ineffectual action, elected Stéphane Dion as their leader in December 2006. It was a rather improbable choice given his high-profile adversaries, Michael Ignatieff and Bob Rae, and Dion's astringent public personality. Dion campaigned, however, on being the greenest of the candidates, outfitted his supporters in green on the morning of the leadership vote, and promised a Canada that would be more "prosperous, fair and green." In Quebec, where he had been lampooned and reviled as a strong defender of Canadian unity, he revived his reputation somewhat through his environmental concerns, since Quebeckers liked to believe themselves excellent environmental custodians, in part because it allowed them to feel morally superior to other Canadians.

Dion, the last Liberal environment minister before the party's defeat, had opposed a carbon tax as a tool for reducing emissions, an idea advanced by Ignatieff. His preference as minister and leadership candidate remained in the world of huge government subsidies, exhortations and voluntarism, the doomed trio of climate change policies. Ignatieff had the right policy, but Dion the better image as an environmentalist.

A year and a half after winning the leadership, Dion swung around to recommending such a carbon tax—the Liberals called it a "green shift"—with revenues from the tax offset by reductions in personal and business taxes. These offsets were heavily weighted towards low- and moderate-income people, which is as any offset policy should be, since they are the people with the least discretionary income, for whom energy is a more significant fixed cost for their personal budget than for wealthier people. The political new positioning also had much to do with Dion's troubled leadership that flowed from his uncharismatic personality but, more importantly, a seeming inability to stake out clear positions, a problem compounded tactically by his unwillingness to force an election that the party believed it could not win. Beset by low polling numbers, and aware of widespread unhappiness with his leadership, Dion searched for a bold initiative that would restore his green credentials, present a clear and (he hoped) compelling alternative to the Conservatives' climate change policy, attract New Democratic Party (NDP) and Green supporters, and finally get Canadians talking about one of his ideas instead of about his weak leadership.

Those were the political considerations motivating the change, but Dion had also evolved intellectually, having watched the introduction of a carbon tax in British Columbia. He came to believe that a "green shift" would over time move people away from fossil fuels more effectively than subsidies or exhortations. He took an enormous political risk, since "green shift" was immediately denounced as a "tax grab" by opponents. The burial grounds of Canadian politics are littered with the gravestones of opposition leaders who laid out bold plans for action: Sir Wilfrid Laurier's trade reciprocity with the United States, Robert Stanfield's wage and price controls, Joe Clark's move of the Canadian embassy in Israel. There were many nervous Liberals who, whatever their concerns about the substance of the "green shift," thought its timing and strategy suicidal, since it offended the first rule of opposition politics, which is to oppose so that an election becomes a vote on the performance of the government, not on the proposals of the Opposition. Predictably, the Conservatives rolled out a television advertising campaign against the Liberal plan even before it was announced. Harper described the Dion plan as "crazy," "insane" and designed to "screw" Canadians.

The NDP remained committed to steep emissions reductions but opposed a carbon tax, thereby disillusioning some of its pro-environment supporters. The NDP argued that a tax would hurt low-income citizens. The party preferred instead to make almost all the burden for reductions fall on large industries through regulations and a cap-and-trade system. For the NDP, taxing carbon was too politically risky an idea for a party determined not to appear to be a "tax-and-spend" left-wing party. The NDP also had links with big industrial unions—steel, auto, paper, chemicals—that feared job losses from action to reduce emissions. The party therefore joined its anti–big business ideology to the fears of its union base, and then added a political calculation about the difficulty of selling a carbon tax. In B.C., the party launched an "axe the tax" campaign against Premier Gordon Campbell's revenue-neutral carbon tax, eyeing seats in the Interior and on Vancouver Island.

The rise in support for the Green Party throughout 2007 and 2008 arguably reflected the biggest change in Canadian party politics. Polls gave the Greens about twice the 5 percent support they had earned in the 2006 election. The party elected a bullhorn of a leader in Elizabeth May, who made a deal with Dion not to run candidates in each other's seats. Her party won 13.4 percent of the vote in March 2008 by-elections in Toronto-Centre and 13.5 percent in Vancouver Quadra. The Greens' vote in the 2008 election jumped to 6.8 percent across the country, an improvement to be sure, but not as large a gain as the party had expected given its sharply higher budget and the publicity surrounding its media-savvy leader, the first Green to participate in a televised leaders' debate.

Arguing against the reality of climate change in polite company became like opposing anti-smoking bylaws, something done by a dwindling number of Canadians, usually furtively, occasionally robustly, always unsuccessfully. Big business went green, at least in advertising and public statements. ExxonMobil, and therefore its Canadian affiliate, Imperial Oil, shifted from funding conferences of climate change deniers to running advertisements promoting its commitment to combating global warming. The Canadian Council of Chief Executives, the big-business lobby group, moved from warning incessantly about the negative economic consequences of serious action against emissions to suggesting that, yes, even a carbon tax might be necessary to curb emissions. The Council was caught, however, between its understanding of how Canadian public opinion had moved and the self-interest of some of its members, for whom any limits to growth or additional environmental burdens should be opposed. The oil and gas industry was also conflicted. After years of successfully avoiding any burdens to reduce emissions, the industry came to understand that its head-in-the-sand approach would no longer work. It also understood, to its surprise, how unpopular the industry was in public opinion, after losing a battle against higher royalties in Alberta. Privately, some senior oil and gas executives preferred a carbon tax to a cap-and-trade system, since the former would apply to every citizen, thereby affirming the industry's old line that everyone, and not just industry, should share in the fight against reducing emissions. But since the federal and provincial Conservatives both opposed a carbon tax, these voices remained mute because they did not want to incur the wrath of governments.

Some of this corporate public shift could fairly be called "greenwash," painting an environmentally friendly face to cover continuing polluting practices; but some of it reflects a response to the swift change in Canadian public opinion. That change altered the political culture of Canada, elevating environmental issues, especially greenhouse gas reductions, to near the top of public priorities and giving the issue an attention it had never experienced, despite the valiant efforts of those who had seen the dangers of climate change and sought, mostly in vain in the 1990s and first half of the

2000–2010 decade, to interest Canadians and alert them to its challenges.

The upward surge in concern about climate change occurred so suddenly that it surprised almost everyone. The surge came after a very long period of solid economic growth, with unemployment at low levels not seen in thirty years. Although climate change critics kept warning that serious action would imperil growth and jobs, the Canadian economy remained so strong once the country emerged from the burdens of deficit and mounting debt in the mid-1990s that the warnings lost some of their audience.

Al Gore's film *An Inconvenient Truth* likely had some effect on Canadians in explaining and dramatizing the global warming phenomenon. In North America, Gore was certainly the best-known exponent of climate change action. His receipt of the Nobel Prize enhanced his profile and reputation. The reports of the Intergovernmental Panel on Climate Change received considerable media coverage. Attention was paid, too, to Sir Nicholas Stern's report about the economic costs of climate change. The Canadian media caught wind of the apparent public shift and began, belatedly, to provide more news and analysis.

Public attitudes, however, do not shift suddenly as a result of reports, media attention and films. The majority of citizens do not pay close attention to public issues. Very few can navigate through the details of any policy. Issues have to be brought to people at a highly symbolic level, or in a way that touches their personal lives, before they ask the most rudimentary questions, let alone form even a marginally informed response, about public issues beyond their immediate lives and preoccupations. Regional variations in opinion invariably mark attitudes towards public issues in Canada. Climate change is no exception, since Canada is cleaved between provinces with substantial deposits of fossil fuels and those without, a fact of geography that, over time, has produced different economies and the political attitudes that flow from them.

Broadly speaking, citizens in provinces with fossil fuels are most hostile to serious action against greenhouse gas emissions, because they fear negative economic consequences, whereas citizens in provinces without fossil fuel industries are most eager for action, at least rhetorically. Any federal government will be caught between these conflicting attitudes. Given these cleavages, the provinces predictably failed to reach consensus on climate change action at a series of interprovincial meetings in 2006, 2007 and 2008. The fossil fuel–producing provinces of Alberta, Saskatchewan and Newfoundland lined up against the rest. The Territories are conflicted internally. Their Arctic geography is the most evidently damaged by global warming, yet their governments fear that economic policies such as carbon taxes or cap-and-trade carbon systems might deter fossil fuel exploration and exploitation in their job-hungry areas.

These internal divisions raise the awkward but pertinent question: how durable and deep was the change in Canadian public opinion towards global warming? Canadian attitudes certainly changed swiftly in the middle of the 2000–2010 decade, causing politicians, business leaders and the media to take the climate change issue more seriously. The awareness of climate change rose sharply, as did the demands for action; while the number of deniers fell precipitously. The nominal willingness of Canadians to see that their governments take serious action jumped, but it remains uncertain, despite the fervent wishes of environmentalists, whether the willingness extends to government policies that produce higher prices for fossil fuels, higher electricity taxes and vehicle costs, and requirements to modify lifestyles and choices in a significant way. The fact that the Liberals played down their "green shift" as the 2008 election went along, and the fact that the party lost seats (albeit for a variety of reasons), suggests a carbon tax remains a non-starter with Canadian voters. Similarly, just a few months after B.C.'s carbon tax went into effect on July 1, 2008, Premier Gordon Campbell's popularity slumped. Both Dion and Campbell were cursed: their tax ideas were made public before the huge jump in oil prices caused the price of gasoline and other fuels to skyrocket. The lesson for politicians and environmentalists was clear: the pain of sharply higher fuel costs considerably outweighed concerns about the long-term effects of climate change.

There remains a very real question—a troubling one for environmentalists—about whether Canadians are anxious to tackle climate change themselves, or whether they prefer that the changes be made by someone else, another region or industry. The NIMBY (Not in My Backyard) syndrome, so evident in municipal planning, might well be at work in climate change politics. By all means, it is said, let's have more wind power, as long as the turbines are not near my property or do not spoil my views. Yes, let us conserve gasoline, but not at the cost of higher prices at the pump. Let us nationally take action, but only if China, India and other countries do likewise. Let us create more public transit, build more trains, reduce garbage, compost more, use electricity less, and engage in all manner of more environmentally sound personal and collective practices—as long as these changes do not cost me more money or result in tax increases. Environmentalists will point to polling data showing Canadians nominally willing to pay more to go green. It is one thing to answer a polling question, but sometimes quite another to pay the actual price. And it is the enduring propensity of groups that want action for a particular cause to read into polling data more support for their cause than reality suggests is warranted.

Environmentalists can kid themselves into translating the upsurge in climate change awareness and concern into automatic support for higher short-term costs to tackle the problem. There are plenty of Canadians—perhaps even a majority—who do not believe in the fiscal neutrality of carbon taxes, because they cannot believe any additional tax money raised by governments will not be wasted. Spending can be by direct outlays or lower taxes (tax expenditures). Either way, Canadians believe that some of this spending will be misdirected or wasted. Whether this perception is correct in fact is politically irrelevant; the perception exists, therefore it is. And every politician knows about the perception-cum-reality, which is why even the New Democratic Party does not call for tax increases of any kind, because it fears accusations of being a "tax-and-spend" party. Statistics Canada has reported that mean family incomes have stagnated for several decades. This stagnation has produced a flinty and understandable reluctance to see more money flowing from households to governments in pursuit of what they see as the public good. Canadians' tax burdens remain, within the Organisation for Economic Co-operation and Development, among the lowest, but Canadians do not live in Europe. They live in North America, adjacent to the United States, their only ongoing measure of comparison. Canadians have definitely gone on green alert.

Canadians are open to ideas about how to tackle greenhouse gas emissions. They want their politicians to get on with the job. But they remain wary of how and at what cost—which is an understandable reaction, since we live in the earliest stages of a long revolution away from such dependence on fossil fuels towards a different kind of energy mix. Although we have long appreciated the advantages of fossil fuels for creating the lifestyles and affluence our societies have enjoyed, we are now more aware that their use brings liabilities in the form of environmental degradation on a world scale. In these early stages of the revolution, there will be policy trials and errors, as governments grope towards policies that work and that can be made politically acceptable. And Canada has particular challenges not faced by, say, European countries, in that Canada's population and economy are growing, its distances are vast, its weather is fiercely cold, its political system is federal, and it possesses vast oil, natural gas and coal reserves, for which there is and will be demand.

A 2008 survey of Canadians' attitudes by the Harris/Decima market research firm, "Understanding Modern

Environmentalism," based on two online surveys of five thousand respondents, found that a strong majority, everywhere in Canada, and across all age groups and party affiliations, felt their generation had done an "unacceptable amount of damage to the environment." The result was a high degree of remorse about the "legacy for future generations," coupled with an openness to remedial action. Whereas it had been conventional wisdom that people eighteen to thirty years old were the keenest for environmental action, the Harris/Decima survey revealed older respondents also worried about their environmental legacy.

Different weather patterns and specific weather-related incidents (storms, floods) led the list of factors that had alerted individuals to the seriousness of climate change. The strange developments touched all parts of Canada. The year 2006 was the second-warmest year on record across Canada, and the tenth consecutive year of above-average temperatures. The winters of 2005 and 2006 were the warmest on record. Winnipeg had its driest June and July ever and its warmest January; Edmonton experienced its highest temperature in seventy years; Toronto's Pearson Airport had its warmest night; Montreal endured its rainiest year. Although few Canadians live in the Far North, some of them heard about the shrinking ice caps and widening of the Northwest Passage. British Columbians could see the sickening work of the mountain pine beetle devastating forests throughout the central and northern parts of the province, the main antidote to which had been very cold winters in previous decades.

The largest number of respondents to the Harris/Decima survey strongly believed that technology would rescue the environment by offering new products and ways of doing things. The most likely changes people believed they would make were the easiest: Energy Star appliances, fluorescent lighting, lowered thermostats; the least likely were changes that would require considerable lifestyle changes, such as driving less, taking public transit more, or using the clothes dryer and dishwasher less. The preferred options for government policy were incentives and subsidies; the least favoured were tax increases.

The Harris/Decima survey did not specifically ask about carbon taxes as they were introduced by Premier Gordon Campbell's British Columbia government and the federal Liberal Party under Dion. In both cases, carbon-producing fuels would be taxed, with the revenues used to lower personal and business taxes. Harris/Decima merely asked about tax increases and, not surprisingly, the response was negative. Which is why opponents of a carbon tax purposefully ignored the second half of the equation and harped only on tax "increases." Still Campbell and Dion deserve the highest marks for political bravery and policy initiative, since they are the first two elected officials in North America to promise or implement carbon taxes.

Federalism, a shaper of Canada's political culture, bedevils and benefits climate change policy. Provinces guard their constitutional authority over natural resources, but only the national government can negotiate internationally for Canada. Both levels of government are responsible constitutionally for the environment. They both have laws and regulations for the environment. Pollution is generated locally, but greenhouse gas emissions do not respect provincial or national boundaries. From wherever they emanate, they enter the atmosphere, the common property of everyone.

Canadian provinces have their own political cultures, and have gone their own ways in developing climate change policies. The Harper government has eschewed any attempt to develop a national policy with the provinces. The prime minister does not like formal federal-provincial conferences, believing correctly that they lead to public posturing and few accomplishments. Any effort to craft a national policy certainly would have been arduous, but the Harper government did not even wish to try. The Harper government preached the virtues of "open federalism," which meant a reluctance to exercise federal leadership in any area that might intrude in whole or in part on provincial jurisdiction. Harper's view is that Ottawa should carry out a more limited number of responsibilities but discharge them effectively, and not intrude, as he believes Liberals were wont to do, in areas of provincial jurisdiction. He is quite willing to let provinces experiment or go their own ways. This attitude towards federalism finds particular approval in Quebec and Alberta.

Any federal attempt at a coordinated federal-provincial approach would necessarily pit Ottawa against Alberta, whose climate change targets are so weak: a 20 percent increase by 2020 and a 14 percent decrease by 2050. The Harper Conservatives with their Alberta base have been unwilling to confront their provincial cousins running the government in Edmonton.

Canada's international negotiators repeat the Harper government's goal for Canada to reduce emissions by 20 percent from 2006 levels by 2020, or only 3 percent from 1990 levels. Climate change talks have been going on for so long that negotiators know each other's positions by heart. Other countries' negotiating teams understand that even Canada's unambitious 20 percent reduction target cannot possibly be achieved if the Alberta government allows emissions to rise by 20 percent by 2020. To do so would require emissions reductions of about 30–35 percent by 2020 in the rest of Canada. Canada, its international reputation already tarnished by having missed Kyoto targets by more than any other signatory to the Protocol, is therefore once again presenting itself on the international stage as a country that says one thing and does another. It is unlikely that Canadians as yet understand this reprise of Canadian international hypocrisy, because the next round of international negotiations are a long way from completion. Polling data has shown, however, that Canadians were somewhat angry when apprised of Canada's tarnished international reputation for climate change inaction. Canadians are, after all, the world's leaders in moral superiority. As the Indigo bookstore chain's slogan puts it: "The World Needs More Canada." Canadians want to believe this slogan, because it suits their self-image as a humanitarian, helpful fixer in the world. Some of the sudden shift in the country's political culture towards climate change resulted from this poor international reputation being relayed back to Canadians. Further international embarrassment would ricochet back into the Canadian political arena.

But for all its many disadvantages, federalism also allows provinces to test policies whose results can be studied by other jurisdictions. When British Columbia became the first jurisdiction in North America to enact a carbon tax, plus a raft of other measures against emissions, Ottawa was content to negotiate an enabling agreement testifying to the compatibility of B.C.'s policy with federal regulations. B.C.'s policy broke the political taboo that no government would dare introduce a carbon tax, given the fraught politics surrounding any tax. But the magnitude of the province's (and country's) challenge was evident in the admission that even the most aggressive set of policies yet devised on the continent would not get the province to its stated target of a one-third reduction by 2020. More still would be required. Whatever good B.C. was doing with these aggressive policies was obliterated by Alberta's approach.

Federalism negated Canada's attaining a coherent national policy; indeed, as in the B.C./Alberta examples, it actually led to policies that cancelled each other out. Eventually, however, the lack of internal coherence will confront certain inescapable international realities. Canada will not be able to defend its international targets if Alberta's are so weak. Alberta runs the real risk of a severe public-relations embarrassment in the United States (and beyond) if its emissions rise while those in the rest of the industrial world fall. Alberta's oil sands are already in the bull's eye of U.S. environmental groups and, increasingly, of environmentally conscious American politicians. The Alberta idea of a provincial-only cap-and-trade system will collapse instantly upon the Americans creating one of their own, as will happen within several years of the

2008 election, at which point every oil and gas CEO in Alberta will be clamouring to join the much wider and more economically efficient U.S. system.

Indeed, the ultimate shock for the Canadian political culture might be to discover that the United States will save Canada from its internal policy incoherence on climate change. Canadians have assumed, usually quite wrongly, that their country is environmentally more advanced than the U.S. Even under President George W. Bush, emissions rose slightly more slowly in the U.S. than in Canada. With the support of U.S. politicians of every stripe, at the national and state level, and with the immense engine of U.S. venture capital now investing heavily in green technologies, the United States is about to leave the Bush years far behind, and adopt tough vehicle emissions standards and a national cap-and-trade system that Canada will have almost no choice but to join, because it would make more economic sense to have tough North American standards than separate and confusing Canadian and U.S. ones. Nothing offends strident Canadian nationalists more than doing something because the Americans are acting and it suits Canada's interests to do likewise. They should prepare themselves now to be offended.

Whatever the vicissitudes of domestic Canadian politics and the vagaries of Canadian public opinion, global warming is a fact. Global warming will not go away. Its effects will worsen, according to the world's best scientific advice. A gap has already widened between the urgency of action and the dilatory Canadian response. That gap bids fair to widen as scientific evidence accumulates about the acceleration of warming and the sputtering, puttering nature of Canadian decisions.

We have learned lessons from recent debates and elections, not all of them pleasant for those who believe in the urgency of action. Yes, the Canadian public is greatly more sensitized to the reality of global warming than, say, five or six years ago. Before action, there must be recognition of a problem. In this sense, Canada is further ahead. But Canadians, it would appear, are deeply conflicted about what kind of action is appropriate, how fast we should turn our attention to it and who should pay. What seems clear is that taxing carbon is politically perilous, even suicidal. Economists can (and should) lament this conclusion; environmentalists might fight against it. But it would appear that for the foreseeable future, taxing carbon has become the third rail of Canadian politics: touch it and you die.

So, Canada will have to develop a range of other approaches that are less frightening to voters, in large part because they will take longer—thereby allowing more time for adjustment—and will cost less, or so the voters believe. These include a cap-and-trade system, now fully supported by the Conservatives (and the NDP), new vehicle emissions programs, and subsidies for renewable fuels, building insulation and research. As the French World War I general said after another costly battle: *il fault reculer pour mieux sauter.*

And that is what those who favour serious action against climate change must do: step back in order to go forward again, with a greater chance of political success. We should not be surprised that this be so, for we are all at the beginning of a long period of industrial, economic and political adjustment towards a world less dependent on fossil fuels. There will be mistakes galore, poor bets, defeated politicians and broken hearts as we grope towards a future we can see but do not fully understand. We can therefore be sobered by what recently happened in Canada, politically speaking, or we can be hopeful that we have learned a lesson, and only fools make the same mistake twice.

It would be desirable if Canada could get its own policies in line, but federalism as described above would seem to preclude coherence. So we will be following the Americans, alas, hoping that in the post-Bush era they will rejoin the international negotiations, commit to serious reduction targets and introduce policies in their country that can be extended to ours, so that together we can make the kind of progress we must make to improve both countries' abysmal records in combating global warming.

Thomas Homer-Dixon and Nick Garrison
CONCLUSION

August 22, 2027

LULISSAT, DENMARK (Reuters)—A steady flow of scientists, journalists, Danish officials and eco-spectators has arrived in this settlement on the west coast of Greenland as thousands of icebergs break off the nearby Jakobshavn glacier each day and stream past the town. In the last four weeks, the glacier has surged off the Greenland land mass, producing a sharp acceleration of ice discharge into the North Atlantic. Glaciers elsewhere on both the east and west coasts of Greenland have shown similar dramatic acceleration of ice loss. Scientists believe these developments signal the start of large-scale disintegration of Greenland's ice sheets and rapid rise of the world's sea levels.

What kind of future are we handing our children? If it includes news stories like the one opposite, today's generation of adults will have failed to meet one of its most fundamental responsibilities—to ensure our children's safety, security and opportunity to flourish. Every day that we deny, dissemble and delay, and every day that we avoid the tough choices and costs of dealing with our twin energy and climate crises, we make it more likely that, at some point, our children, perhaps as young adults, will wake up in a world in which they no longer control their climate fate—in which, for instance, it's too late to stop the collapse of the planet's great ice sheets.

Although the authors of this book's chapters disagree about many things, they all agree, fundamentally, that we need to change how we see the world and how we act. In our every-day lives, we need to see and understand the energy and climate consequences of our behaviour, and we need to change that behaviour to protect our children's future. All the authors here are deeply frustrated by the half-heartedness of our efforts so far and the current poverty of serious proposals to alter our direction. And they all fear we won't take these challenges seriously until it's too late.

In the period during which this book was written and edited, the world seemed to go haywire. An economic bubble created by decades of cheap credit and the undervaluing of risk, centred in the U.S. housing sector (but replicated in different ways in economies around the world), imploded, producing a financial crisis of astonishing breadth, depth and persistence. Commodity prices that had rocketed upward suddenly collapsed as demand vanished and hedge funds and other speculators sold their positions to pay down their debts. Oil, which spiked to $147 a barrel in June 2008, plummeted below $60 by mid-November 2008.

But almost unnoticed in the background of this tumult, companies and utilities were ramping up their use of coal, and humankind's emissions of carbon dioxide shot past the worst-case scenario of the Intergovernmental Panel on Climate Change. In the summers of 2007 and 2008, Arctic ice disappeared at record speed. Meanwhile, in British Columbia and in federal politics, Canadians showed scant enthusiasm for proposals to tax carbon emissions.

When it comes to our energy and climate problems, we're going in diametrically the opposite direction to where we should be going. And while the world's current economic crisis legitimately demands policy-makers' attention, its consequences, no matter how severe, will pale beside the long-term consequences to humankind of energy scarcity and climate change.

For all their capacity to radically alter nearly every facet of our civilization, these threats rarely capture our imaginations. They develop incrementally and their consequences are both uncertain and relatively distant in time and space.

They are what specialists call "wicked problems." As David Keith points out in his chapter here, climate change, in particular, is characterized by high uncertainty and high inertia. Systems with lots of uncertainty and inertia are notoriously hard to control, because we can't effectively predict their future behaviour, and we can't quickly correct behaviour we don't like. So such systems give us enormous scope for denial and procrastination.

But the carbon challenge is real, and deep change in our lives, economies and societies is inevitable. It would be a mistake to conclude from the disagreements between the experts in this book that we should wait to see what happens before we act. David Keith and David Hughes may argue about the relative importance of climate change and fossil fuel depletion. Mark Jaccard and Jeff Rubin may disagree on whether markets will accurately price the energy that powers our civilizations. But to interpret these disagreements as reasons to delay would be a self-serving rationalization, just as self-serving as the commonplace rationalizations highlighted by William Marsden and Jeffrey Simpson—that we can enjoy the wealth and energy of the tar sands without acknowledging the staggering environmental costs, and that we can soothe our consciences by endorsing a "green" political agenda, even as we vote for parties and policies that fail to address the problems we claim we want solved.

All the authors in this book agree we face a series of challenges that will shake our priorities, our assumptions about our entitlements and our expectations for the future. Whether the risk is that humankind will release enough of Earth's stored carbon to make the planet uninhabitable, as David

Keith fears, or that humankind will find there aren't enough hydrocarbons in the ground to keep the global economy operational, as David Hughes warns, a carbon shift—either voluntary or involuntary—is now unavoidable.

It had better be voluntary. To do little or nothing now in the face of inevitable change is to invite disaster, if not upon ourselves, then upon future generations. In fact, one way or the other, whether we change our behaviour or not, we're betting the fate of our children, and the simple question is: Which bet do we want to make? Do we bet that energy scarcity and climate change aren't going to hurt the world badly, and invest our resources elsewhere?

If we're right, we save some money, but if we're wrong, the consequences for our children could be catastrophic.

Or do we bet that energy scarcity and climate change could indeed hurt the world badly, and invest to prevent that outcome?

If we're right, our children avoid possible catastrophe, but if we're wrong, we lose some money.[1]

Put this way, the answer seems obvious: investment now to prevent potentially catastrophic consequences in the future, even very significant investment, is entirely rational. Or at least the answer should seem obvious, if we're not deeply hardened in our narrow, narcissistic, present-day self-interest.

But logic alone isn't going to cause us to act. Some kind of sharp shock or sudden crisis will probably be needed to galvanize public opinion and mobilize financial and political capital to confront the energy–climate threat. Hurricane Katrina was only a Category 3 storm, and though the damage and misery it caused were compounded by authorities' failures, a more powerful storm slamming into another coastal city could do far more damage. The question isn't really whether something like this will happen, but whether it will have a mobilizing effect and whether that effect will be salutary. The attacks of September 11 focused public attention on "terror" and furnished much of the world, and especially the United States, with a new vocabulary and a new set of shared and politically inviolable ambitions—no politician could afford to be "soft" on terror. We now know that this vocabulary and these shared ambitions were distorted and abused by the Bush administration—in part to justify the invasion of Iraq. If our societies won't act decisively to address the energy–climate challenge until a crisis hits, then we must take care that any post-crisis crystallization of public opinion isn't abused in similar ways.

Once humankind decides that real action is necessary, the task of refitting our global civilization to run on a carbon-free energy will be heroic in scale. Very few people grasp its true magnitude. Consider the oft-touted hydrogen option. To replace the oil burned in today's American fleet of 230 million cars, trucks and buses would require 230,000 tonnes of hydrogen each day, enough to fill 13,000 *Hindenburg* dirigibles. To produce this hydrogen by electrolyzing water, the U.S. would have to nearly double its total electricity generating capacity. And to generate this electricity from carbon-free renewable sources would require covering an area the size of Massachusetts with solar panels or of New York State with windmills.[2] The United States would also have to pay the staggering costs of building the pipeline system to distribute the hydrogen, the filling stations to deliver it and the new fleet of vehicles to use it. The mind boggles.

We're going to have to get used to mind-boggling changes, because the kind of undertaking described in the foregoing paragraph will be only one component of our transition to a sustainable future. The starting point of this transition will almost certainly be an economically significant, and incrementally rising, price on carbon emissions. Once such a price is established—whether it's a product of a carbon tax or a cap-and-trade system—the entrepreneurial energies of capitalism will shift to focus on keeping every kilogram of carbon out of the atmosphere. We'll adopt (by current standards) stunningly ambitious measures for energy efficiency and conservation. And because coal will likely take centre stage as oil supplies tighten (especially the conversion of coal to transportation fuel), we'll deploy, on an enormous scale, technologies to pump the carbon dioxide produced by using coal underground.

We'll also need to invest huge resources in scientific research and development to find potentially game-changing energy technologies, like underground coal gasification and enhanced geothermal power generation. (The former involves producing a gaseous hydrocarbon mixture in deep, unmineable coal seams, bringing the mixture to the surface, stripping out the hydrogen and then pumping the residual carbon dioxide back underground; the latter involves drilling several kilometres into Earth's crust, pumping water down the hole, where it's heated to hundreds of degrees Celsius, and then bringing the superhot water back to the surface to drive electrical turbines.)

Moreover, some leading climate scientists are now arguing that we need to not only reverse the steady rise of global carbon *emissions* and ramp them down to zero, but that we also need to ramp down the *level* of carbon in the atmosphere. One of the most significant recent advances in climate science has been much better understanding of the relative balance between positive (self-reinforcing) and negative (self-equilibrating) feedbacks. Earth's climate system, both its physical and biological components, contains many feedbacks, and research now indicates that the climate system's positive feedbacks outnumber and, in their aggregate force, strongly outweigh its negative feedbacks. As James Hansen of the Goddard Institute at NASA and his colleagues have recently written in a top, peer-reviewed journal:

> Palaeoclimate data show that the Earth's climate is remarkably sensitive to global forcings. Positive feedbacks predominate. This allows the entire planet to be whipsawed between climate states . . . Recent greenhouse gas emissions place the Earth perilously close to dramatic climate change that could run out of our control, with great dangers for humans and other creatures.[3]

A decade ago, scientists generally believed that the critical threshold at which Earth's climate might start warming on its own accord—that is, the point at which positive feedbacks would take over and the biosphere would start releasing carbon into the atmosphere (from melting permafrost, for instance) in quantities large enough to create a self-reinforcing cycle—was in excess of a doubling of pre-industrial levels of atmospheric carbon dioxide, or above 560 parts per million.

More recently the scientific consensus shifted, and the critical level was widely defined as about 450 parts per million. But in the last couple of years, some climate scientists like Hansen, after analyzing the latest data on the climate system and biosphere's responses to warming, have concluded that the current concentration of around 385 parts per million (rising between 2.0 and 2.5 parts per million a year) is already well over the red line. So they've been suggesting that humankind should aim for a target of 350 parts per million.

At a time when global emissions of carbon are soaring relentlessly upward, it seems almost crazy to suggest we should be aiming not only to turn the trend around and ramp emissions down to zero—something that in itself seems virtually impossible in a world addicted to fossil energy—but actually to *lower* atmospheric concentrations. Yet the data on climate change are as relentless as humankind's rising emissions, and these data are telling us that Earth's climate appears to be far more sensitive to changes in the atmosphere's level of greenhouse gases than was understood previously. And impossible though it may seem, at a carbon price above eighty to one hundred dollars per tonne, it would likely be profitable to extract carbon dioxide from the atmosphere and pump it underground, assuming a carbon-free energy source powers the process.

Even with such heroic interventions, however, the atmosphere's levels of carbon dioxide will almost certainly continue to rise for several decades—and perhaps for much of this century. We'll steadily approach that unknown and perhaps unknowable threshold at which warming becomes its own cause. To head off runaway climate change, and to prevent catastrophic melting of the world's great ice sheets in Greenland and the Antarctic, we may have to intervene intentionally in the global climate to reduce warming, especially at the planet's poles. Such "geoengineering" could, for example, involve injecting large quantities of sulfate aerosols into the polar stratosphere to make the atmosphere reflect more incoming solar radiation.

It's an appalling prospect—a hubristic intervention in a complex system that's incompletely understood, and an intervention that could easily backfire. But the world's publics, when confronted with news like the fictional wire story at the start of this chapter, may demand immediate action. Faced with an existential threat and an urgent clamour for something to be done, our societies could move to a state akin to a war footing, as they pivot to focus their human and financial capital on the energy–climate challenge. In the past, such emergency mobilizations of social resources and savings have often been accompanied by draconian restrictions on human rights and freedoms.

In the end, when we finally face the magnitude of our energy–climate challenge and decide to act, the most difficult change we'll have to make may be the most fundamental: curtailing global economic growth. Over the past century, the world's great ideological systems have disagreed about many things, but the importance of endless economic growth hasn't been one of them. During the twentieth century, communism and capitalist democracy confronted each other around the globe, but neither questioned the growth imperative, because the political and social stability of both systems depended fundamentally on growth.

Today, for many people in poor societies, economic growth often means the difference between bare survival and some modest measure of economic well-being; in the world's poorest societies, growth can mean the difference between death and life. Within one human lifetime, rapid growth has lifted countries like South Korea out of agrarian poverty, and it holds the promise of doing the same for the rest of the world's poor.

Surprisingly, rich societies are just as dependent on endless growth. In much the same way that we need to keep discovering and developing new oil and gas wells at an ambitious pace just to maintain current production, wealthy economies need to grow at between 3 and 5 percent each year—doubling every fifteen to twenty-five years—just to keep unemployment from rising and maintain social peace. The Great Depression showed that a seemingly enduring social contract can quickly erode into political extremism when consumer demand fails, the economy contracts and unemployment soars. Without growth, competition between rich and poor becomes a zero-sum game. A world without growth is not just bad news for investors—it's politically and socially explosive.

And yet, like the wealth-generating waters of the Pactolus, growth appears to be as much a curse as a blessing. Humankind faces a basic, and largely unacknowledged, paradox: infinite growth is necessary for our well-being, but it's impossible in a world of finite resources, and if we pursue it, we'll wreck our well-being anyway. Like Midas, if we try to turn everything into gold, we will be left with an uninhabitable world.

Growth—today defined as the manufacture and consumption of ever more stuff—means the consumption of ever-greater amounts of the energy and raw materials that constitute that stuff. There are those who claim that our increasingly sophisticated information economies hold out the promise of diminishing resource dependence and steadily declining energy and material use per dollar of gross domestic product. Although it's true that, since the 1970s, rich countries have made great improvements in energy and resource efficiency, claims that we can perpetually grow simply by steadily improving efficiency overlook the fact that investments in efficiency show diminishing returns—the easiest things are done first, and the most expensive are left for last. Efficiency can't be improved at high rates indefinitely, because eventually the costs outrun the benefits.

Also, the vaunted improvements in rich countries' efficiency in recent decades have been accomplished in significant part by moving production to countries whose pollution, including carbon emissions, we're now complaining about. We may use less energy and raw materials and pollute less per unit of gross domestic product, but that's largely because it's cheaper to consume things made in China than it is to buy the same thing made in North America. We've exported our pollution and appetite for resources to the other side of the world.

There are those, of course, who argue strenuously and with great acumen that growth is not only essential but can also be perpetual. In a world imperilled by energy scarcity and climate change, though, these people now bear the burden of proof. They need to explain precisely how today's seventy-trillion-dollar world economy—an economy already placing enormous strain on the world's resources and ecological and climate systems—can quadruple in size in real terms in the next fifty years (which it will, if we hold to recent rates of growth) without destroying those vital systems.

Again, voluntarily or not, growth as currently defined is going to stop this century, and we'll all be a lot better off if the process is voluntary. If business as usual simply can't continue, we'd better start thinking hard about what our next sort of business might look like.

All the above ideas about what the future will hold, both for us and for those in the next generations dear to us, may seem like science fiction—particularly when the daily world around us still doesn't look all that different from the way it did yesterday. We still drive to work and to our kids' hockey and soccer games; we still eat and live and vacation as we have for years. It's hard to imagine that these regularities of our lives will change, and perhaps change abruptly.

But the Soviet Union looked every bit as permanent in 1985 as our civilization does today. Its collapse a few years later caught the world by surprise, and we learned that even seemingly powerful institutions—and even seemingly entrenched patterns of life—can be fragile and ephemeral.

While few people grasped in advance the true gravity of the stresses that ultimately caused the Soviet empire's demise, we don't have the same excuse today. We understand the energy, climate and other challenges we face well enough to know that ignoring them could be catastrophic. We also understand, for the most part, what we need to do to avoid catastrophe.

A carbon shift is coming. We can see it in front of us. If we continue to deny, delay and dissemble—and simply hand the energy–climate problem, unsolved and likely vastly worse, to our children—they will have no one but us to blame.

Thomas Homer-Dixon and Nick Garrison

Notes

Introduction | Thomas Homer-Dixon and Nick Garrison

1. Dominic Wilson and Roopa Purushothaman, "Dreaming With the BRICs: The Path to 2050," *Global Economics Paper,* no. 99 (October 2003): 4.

2. Jeremy Leggett, *The Empty Tank: Oil, Gas, Hot Air, and the Coming Global Financial Catastrophe* (New York: Random House, 2005): xi.

3. Thomas Homer-Dixon, *The Upside of Down: Catastrophe, Creativity, and the Renewal of Civilization* (Toronto: Knopf Canada, 2006): 83.

4. Robert Solow, "A Contribution to the Theory of Economic Growth," *Quarterly Journal of Economics* 70, no. 1 (65–94): 1956.

5. Reiner Kümmel, et al., "Capital, labour, energy and creativity: Modeling innovation diffusion," *Structural Change and Economic Dynamics* 13 (2002): 415–33.

6. Ibid, 418.

7. Robert U. Ayres and Benjamin Warr, "Accounting for Growth: The Role of Physical Work," *Structural Change and Economic Dynamics* 16 (2005): 181–209.

8. National Petroleum Council, *Facing the Hard Truths about Energy: A Comprehensive View to 2030 of Global Oil and Natural Gas,* draft report (July 18, 2007): 7.

9. David Strahan, *The Last Oil Shock: A Survival Guide to the Imminent Extinction of Petroleum Man* (London: John Murray, 2007): 23.

10. National Petroleum Council, *Facing the Hard Truths about Energy,* 8.

11. Strahan, *The Last Oil Shock,* 114.

12. International Energy Agency, "Analysis of the Impact of High Oil Prices on the Global Economy" (May 2004): 2.

13. Robert L. Hirsh et al., *Peaking of World Oil Production: Impacts, Mitigation, and Risk Management* (United States Department of Energy, February 2005): 31.

14. Jeffry Currie et al., "The Sustainability of High Oil Prices: Revenge of the Old Economy, Part 2" (Goldman Sachs Commodity Research, June 8, 2004): 1.

15. James Lovelock, *The Revenge of Gaia: Why the Earth Is Fighting Back— and How We Can Still Save Humanity* (London: Allen Lane, 2006): 45.

16. Munich Re 1997, Annual Review of Natural Catastrophes 1996, Munich: Munich Re.

17. Kurt M. Campbell, Alexander T. J. Lennon, Julianne Smith, et al. *The Age of Consequences: The Foreign Policy and National Security Implications of Global Climate Change* (Center for Strategic and International Studies and Center for a New American Security, November 2007): 55.

18. Institute of Physics, "Greenhouse Gas Emissions Set to Rise as New Sources for Transport Fuel Are Used" (December 7, 2006), *http://www.iop.org/News/Community_News_Archive/2006/news_9600.html* (accessed November 17, 2008).

19. Natural Resources Defence Council, "Why Liquid Coal Is Not a Viable Option to Move America Beyond Oil," February 2007, *http://www.nrdc.org/globalWarming/coal/liquids.pdf* (accessed November 17, 2008).

Dangerous Abundance | David Keith

1. Estimates of future resources are highly uncertain. One could credibly defend estimates several times smaller than those I used here as well as estimates that are significantly larger. These estimates are drawn

from the studies IIASA/WEC world energy studies (while there are more recent versions, the following is a good summary: Rogner, H-H, *An assessment of world hydrocarbon resources,* Annual Review of Energy and the Environment, 22: 217–262, 1977). I used the studies because they use reasonably transparent methodology and they attempt to take systematic account of the technological and economic drivers that convert resources into reserves.

2. Estimates of resource scarcity often discount technological change that works to facilitate access to resources even as the average quality of the resource declines. Discussions of coal scarcity, for example, typically do not even consider advanced autonomous mining technologies or underground gasification, yet underground gasification has already been used successfully in several commercial or near-commercial demonstrations. Likewise, arguments for scarcity typically discount the possibility of accessing methane hydrates on the seabed. The most radical claims about scarcity of fossil resources implicitly depend on the assumption that a whole set of technologies—any one of which could greatly expand the reserve base—will each separately fail in the next century. While no sensible person expects all these technologies to succeed, the probability of them all failing seems vanishingly small. Underestimating technological change is the underlying reason why so many historical claims about resource scarcity have proved false over the last century and a half.

3. Perhaps the strongest arguments for coal scarcity are made by David Rutledge, a professor of electrical engineering at Caltech. Most of the arguments rest on fitting historical data to a logistic production curve (see http://rutledge.caltech.edu). This method might be convincing when one is well past the peak of production, but prior to the peak the method interprets any downward deviation of production from a logistic/exponential as a reduction in the ultimate resource. This might

make sense if there was a convincing reason that the production rates should "naturally" follow an exponential growth trajectory and if any deviation from exponential growth was due to resource scarcity. If scarcity were playing a role one would expect prices to be rising, but in fact U.S. coal prices have remained remarkably constant (aside from any peak in the mid-70s) since the 1950s during which time annual production has more than doubled. This price history contradicts the assumption that resource scarcity is driving the production rate. Far more plausibly, other factors in the U.S. electricity markets have caused production to increase linearly rather than exponentially, factors that have little to do with coal price or scarcity. If coal production rates have had little to do with coal prices and scarcity, then there is no reason to expect that the logistic fit would mean anything, even assuming that there was sufficient data to make a high-quality fit (which there is not).

The Energy Issue: A More Urgent Problem than Climate Change? | J. David Hughes

1. EIA International Energy Outlook 2008—reference case, *http://www.eia.doe.gov/oiaf/ieo/ieorefcase.html*.

2. OECD/IEA, *World Energy Outlook 2008* — *http://www.worldenergyoutlook.org/weo2008/weo2008-es-english.pdf*: 42.

3. EIA International Energy Outlook 2008—reference case, *http://www.eia.doe.gov/oiaf/ieo/ieorefcase.html*.

4. OECD/IEA, *World Energy Outlook 2007*, Part 1: 43, 64.

5. EIA International Energy Outlook 2007—reference case, *http://www.eia.doe.gov/oiaf/archive/ieo07/index.html*.

6. OECD/IEA, *World Energy Outlook 2007*, Part 1: 80.

7. EIA International Energy Outlook 2008—reference case, *http://www.eia.doe.gov/oiaf/ieo/ieorefcase.html*.

8. *http://www.blacksunjournal.com/energy-transition/62_todays-oil-is -obscenely-cheap_2005.html.*

9. C. J. Campbell, personal communication with the author (October 2007).

10. R. L. Hirsch, "Peaking of World Production: Recent Forecasts," *DOE/NETL Report 2007/1263*, 21.

11. *http://www.nytimes.com/2008/05/29/business/29gas.html?_r=1&oref=slogin.*

12. *http://www.platts.com/Natural%20Gas/Resources/News%20Features/ globallng08/index.xml.*

13. *ftp://ftp.nerc.com/pub/sys/all_updl/docs/pubs/LTRA2006.pdf.*

14. *http://www.neb.gc.ca/clf-nsi/rnrgynfmtn/nrgyrprt/nrgyftr/ 2007/nrgyftr2007-eng.html.*

15. *ftp://ftp.nerc.com/pub/sys/all_updl/docs/pubs/LTRA2006.pdf.*

16. *http://www.energywatchgroup.org/fileadmin/global/pdf/EWG_Report_Coal _10–07–2007ms.pdf.*

17. *http://dels.nas.edu/dels/rpt_briefs/coal_r&d_final.pdf.*

18. *http://www.tsl.uu.se/uhdsg/Publications/Coalarticle.pdf.*

19. *http://rutledge.caltech.edu/.*

20. *http://www.census.gov/ipc/www/idb/worldpop.html.*

21. Lindsey Grant, *The Collapsing Bubble: Growth and Fossil Energy* (Santa Ana, CA: Seven Locks Press): 74.

22. *http://www.npg.org/forum_series/tightening_conflict.htm.*

23. *http://www.upi.com/Emerging_Threats/2008/02/27/Walkers_World_The _coming_food_crisis/UPI-60181204127086/.*

24. *http://www.nationalpost.com/news/story.html?id=412984.*

25. Vaclav Smil, *Energy in Nature and Society: General Energetics of Complex Systems* (2008).

Peak Oil and Market Feedbacks: Chicken Little versus Dr. Pangloss | Mark Jaccard

1. M. Jaccard, *Sustainable Fossil Fuels: The Unusual Suspect in the Quest for Clean and Enduring Energy* (Cambridge, UK: Cambridge University Press, 2005).

2. One tonne of oil equivalent equals forty-two gigajoules; one thousand cubic metres of natural gas equals one gigajoule; and one tonne of coal equivalent equals twenty-nine gigajoules on a thermal basis. See H.-H. Rogner (lead author Chapter 5) in J. Goldemberg, ed., *World Energy Assessment: Energy and the Challenge of Sustainability* (New York: United Nations Development Programme, 2000): 139.

3. His earlier analysis is summarized in M. Hubbert, "National Academy of Sciences Report on Energy Resources: Reply," *Bulletin of the American Association of Petroleum Geologists* 49, no. 10 (1965): 1720–27.

4. C. Campbell and J. Laherrère, "The End of Cheap Oil," *Scientific American* 278, no. 3 (1998): 78–84; R. Bentley, "Global Oil and Gas Depletion: An Overview," *Energy Policy* 30, no. 3 (2002): 189–205; and K. Deffeyes, *Hubbert's Peak: The Impending World Oil Shortage* (Princeton, NJ: Princeton University Press, 2001).

5. E. Zimmerman, *World Resources and Industries* (revised ed.) (New York: Harper and Brothers, 1951): 15.

6. For recent efforts by economists to model peak oil, see the following two articles. S. Holland, "Modeling peak oil," *The Energy Journal* 29, no. 2 (2008): 61–80. A. Brandt, "Testing Hubbert," *Energy Policy* 35 (2007): 3074–88.

7. For example, see P. Odell, *Why Carbon Fuels Will Dominate the 21st Century's Global Energy Economy* (London: Multi-Science, 2004); M. Adelman and G. C. Watkins, "Reserve prices and mineral resource theory," *The Energy Journal*, Special issue (2008): 1–16; and S. Holland, "Modeling peak oil," *The Energy Journal* 29, no. 2 (2008): 61–80.

8. The figure is from C. Cleveland and R. Kaufmann (2008), "Fundamental

principles of energy," in *Encyclopedia of Earth*, ed. Cutler J. Cleveland (Washington, DC: Environmental Information Coalition, National Council for Science and the Environment). [First published in *Encyclopedia of Earth* April 1, 2008; last revised April 22, 2008; retrieved April 25, 2008.]

The Perfect Moment | William Marsden

1. The Alberta Energy and Utilities Board claims that the tar sands contain approximately 1.7 trillion barrels of bitumen, of which about 174 billion barrels are recoverable with current technology, and 315 billion barrels will be recoverable, it is hoped, with technological advances. If these estimates are true, this makes the tar sands the largest petroleum resource in the world.

2. And the solution to the depletion problem is not the fabled energy riches under the melting Arctic ice. Geologists estimate that the Arctic holds about 90 billion barrels of oil. At present consumption this will last the world about three years.

3. Environment Canada, "Greenhouse Gas Sources and Sinks in Canada," *National Inventory Report 2008*, 77.

4. Ibid.

5. Richard Cairney, Oil Sands Tailings Research Facility launch (University of Alberta 2004), opening statement.

6. Kevin P. Timoney, *A Study of Water and Sediment Quality as Related to Public Health Issues* (Fort Chipewyan, AB), 4.

7. The potential size of these forest carbon emissions is indicated by what is happening in British Columbia. Scientists at the B.C. forestry department warn in a 2008 report that the pine beetle infestation, which has an unprecedented intensity that they attribute to global warming, is turning the forest into a carbon emitter. Carbon emissions

from decaying trees—predicted to total 270 million tonnes from 2000 to 2020, or 38 percent of Canada's annual total emissions—are not included in Canada's greenhouse gas emission models, the scientists note.

8. Environment Canada, "Greenhouse Gas Sources," 15–16.

9. Suncor Progress Report on Climate Change, *http://www.suncor.com/data/1/rec_docs/1752_cc_report_2008.pdf*, 3.

10. Syncrude, "Production Since Startup July, 1978," *http://www.syncrude.ca/users/folder.asp*. Syncrude's website counter records the ongoing production of Syncrude Sweet Blend.

11. U.S. Energy Information Administration, *Energy Consumption by Sector, Selected Years, 1949–2007*, table, *Annual Energy Review 2007*, *http://www.eia.doe.gov/emeu/aer/pdf/pages/sec2_4.pdf*.

12. *http://www.energy.gov.ab.ca/*.

13. Figures from U.S. Environmental Protection Agency; U.S.-based Environmental Integrity Project and Canada's Environmental Defence report (June 5, 2008).

14. *www.finance.alberta.ca/business/ahstf/index.html*.

15. The October 2008 financial crisis has reduced the stock price of major Canadian energy companies by more than 50 percent. Suncor Inc.'s market capitalization, for example, fell to $27 billion from $70 billion. The crisis offered a good opportunity for Canadians to nationalize the vital sectors of their industry and take back control of their energy and environmental future.

16. You could argue that you can sell a natural system such as a waterfall. But in fact what you are selling is its potential as, say, a hydroelectric producer, a tourist attraction or a pleasing view that simply enhances the value of real estate. All of these conceptions would inevitably lead to alterations that would profoundly change the natural system, which would still continue to have no intrinsic value.

Conclusion | Thomas Homer-Dixon and Nick Garrison

1. Some readers will recognize this as a version of the seventeenth century French mathematician and philosopher Blaise Pascal's famous wager regarding the existence of God.

2. Paul Grant, "Hydrogen Lifts Off—With a Heavy Load," *Nature* 424, no. 6945 (July 10, 2003): 129–30.

3. James Hansen, et al. "Climate and Trace Gases," Phil. Trans. R. Soc. A 365 (2007): 1925-54.

4. Benjamin Friedman, *The Moral Consequences of Economic Growth* (New York: Knopf, 2005).

gasoline consumption and, 135–37

peak production of. *See* peak oil

political controls over the supply of, 14, 77–78, 138–47

prices of, 10–14, 52–53, 61–62, 98–100, 115–17, 120–31, 133–51

production rates of, 9–13, 62–63, 76–78, 143–49

refineries and the supply of, 13, 78

reserves of, 9–10, 28, 59–63, 74, 97, 100–107, 112–13, 155, 157

tankers for transport of, 13

types of defined, 105–7

oil sands. *See also* tar sands (Alberta's), 31, 34, 68, 72, 74, 100, 106, 114

extraction requirements in, 156–57

oil shales. *See also* oil sands, 12, 68, 74, 76, 78, 81–82, 100, 106–7, 113–16, 157

Ontario, 167, 183

OPEC. *See primarily member countries e.g.* Saudi Arabia, 9, 133–35, 138–49, 150

peak food production, 90

peak oil. *See also* oil, 9–11, 75–77, 97, 101

shifting definitions of, 110–14

permafrost, melting of, 16

pollution (atmospheric), 51, 53–54, 179–80

population (human), *vii–ix*, 5

"progress traps," *ix*

Qatar, 114

Rae, Bob, 184

rationing, of fuel, 139

Raymond, Lee, 8

Reform Party (Canada), 179, 183

renewable energy (sources and technology). *See also* biomass; energy, sustainable; geothermal power; hydrogen power; hydro power; solar power; wind power, 46, 64–67, 71, 85, 91–93, 105, 118, 131, 200, 207

reserves (fuel). *See also fuel types, e.g.* oil, reserves of, 9, 12, 30–32, 69, 104–10

defined, 30–31, 103, 106

resources (fuel). *See also fuel types, e.g.* natural gas; oil, 4, 9, 67, 69–74, 89, 104–10, 112–13, 116, 213–14

defined, 30–32, 103, 106, 112–13

revolutions (political). *See* unrest (civil)

Rubin, Jeff, 21, 24, 132, 205

Russia, 3, 11, 14, 19, 21, 76, 79, 105, 136–37, 144–45, 150, 214

oil production and reserves in, 146–49

Saskatchewan, 190

Saudi Arabia, 10–12, 60, 67–68, 77–78, 106, 145–46, 156

oil prices within, 138–44

sea level, changes in, 17–18, 41–42

security (national)

climate change and, 18–19

energy scarcity and, 89

shale. *See* oil shales

Simpson, Jeffrey, 24–25, 176, 205

solar power

advantages of, 47

drawbacks associated with, 46, 71

Solow, Robert, 6–7

South Africa, 114, 117

Soviet Union. *See* Russia

Stanfield, Robert, 186

Stelmach, Ed, 153

Stern, Sir Nicholas, 18, 189